HD
1698
M4
W38
2011

Water Rights and Social Justice in the Mekong Region

Water Rights and Social Justice in the Mekong Region

EDITED BY

*Kate Lazarus, Nathan Badenoch, Nga Dao
and Bernadette P. Resurreccion*

London • Washington, DC

First published in 2011 by Earthscan

Copyright © Unit for Social and Environmental Research (USER), Chiang Mai University, Thailand 2011

All rights reserved. No part of this publication may be reproduced, stored in a retrieval system, or transmitted, in any form or by any means, electronic, mechanical, photocopying, recording or otherwise, except as expressly permitted by law, without the prior, written permission of the publisher.

Earthscan Ltd, Dunstan House, 14a St Cross Street, London EC1N 8XA, UK
Earthscan LLC, 1616 P Street, NW, Washington, DC 20036, USA

Earthscan publishes in association with the International Institute for Environment and Development

For more information on Earthscan publications, see www.earthscan.co.uk or write to earthinfo@earthscan.co.uk

ISBN: 978-1-84971-188-3 hardback

Typeset by JS Typesetting Ltd, Porthcawl, Mid Glamorgan
Cover design by Susanne Harris

A catalogue record for this book is available from the British Library

Library of Congress Cataloging-in-Publication Data

Water rights and social justice in the Mekong region / edited by Kate Lazarus ... [et al.].
     p. cm.
  Includes bibliographical references and index.
  ISBN 978-1-84971-188-3 (hbk.)
 1. Water resources development–Mekong River Region. 2. Water-supply–Mekong River Region–Management. 3. Water rights–Mekong River Region. 4. Right to water–Mekong River Region. 5. Social justice–Mekong River Region. I. Lazarus, Kate.
  HD1698.M4W38 2010
  333.9100959–dc22
                              2010039473

At Earthscan we strive to minimize our environmental impacts and carbon footprint through reducing waste, recycling and offsetting our CO_2 emissions, including those created through publication of this book. For more details of our environmental policy, see www.earthscan.co.uk.

Printed and bound in the UK by T. J. International, an ISO 14001 accredited company. The paper used is FSC certified and the inks are vegetable based.

Contents

Figures, Tables and Boxes   *vii*
Contributors   *ix*
Preface: About M-POWER   *xiii*
Acknowledgements   *xv*
Acronyms and Abbreviations   *xvii*

1   Water Governance and Water Rights in the Mekong Region   1
    Nathan Badenoch, Kate Lazarus, Bernadette P. Resurreccion and Nga Dao

PART I – PARTICIPATION IN DECISION-MAKING

2   Water Transfer Planning in Northeast Thailand: Rhetoric and Practice   19
    Philippe Floch and David Blake

3   Local People's Participation in Involuntary Resettlement in Vietnam:
    A Case Study of the Son La Hydropower Project   39
    Tran Van Ha

PART II – SOCIAL DIFFERENCES AND ACCESS

4   Rights and Rites: Local Strategies to Manage Competition for
    Watershed Resources in Northern Thailand   67
    Nathan Badenoch and Prasit Leepreecha

5   Local Institutions and the Politics of Watershed Management in the
    Uplands of Northern Thailand   91
    Rajesh Daniel and Songphonsak Ratanawilailak

6   Gender, Commercialization and the Fisheries–Aquaculture Divide in
    the Mekong Region   115
    Louis Lebel, Santita Ganjanapan, Phimphakan Lebel, Mith Somountha,
    Tran Tri Ngoc Trinh, Geeta Bhatrai Bastakoti and Chanagun Chitmanat

PART III – COMPETING DEMANDS AND PROTECTING THE RIGHTS OF THE MARGINALIZED

7 Fisheries, Nutrition and Regional Development Pathways: Reasserting Food Rights 149
   Robert Arthur, Richard Friend and Mark Dubois

8 Livelihood and Environment Trade-offs at the Time of *Doi Moi*: Industrial Water Use and Wastewater Management in a Craft Village in Peri-urban Hanoi 167
   Le Thi Van Hue and Edsel E. Sajor

PART IV – CLIMATE CHANGE AND THE RIGHTS OF THE VULNERABLE

9 Climate Change in the Asian Highlands: Socio-economic Implications for the Mekong Region 197
   Jianchu Xu

10 Linking Climate Change Risks and Rights of Upland Peoples in the Mekong 217
    Jianchu Xu and Rajesh Daniel

PART V – CONCLUSION

11 Ensuring Justice in Water Governance in the Mekong Region 245
    Bernadette P. Resurreccion, Nga Dao, Kate Lazarus and Nathan Badenoch

Index 255

Figures, Tables and Boxes

FIGURES

2.1	Option for Nam Ngum–Northeast Thailand water transfer	28
3.1	Map of Son La Dam location and field study sites	47
4.1	Rites and rights in watershed management	72
4.2	Huai Sai Khao irrigation	82
5.1	Local politics of watershed management	93
6.1	Transitions from capture to culture or commercialization may affect gender relations and vice-versa	117
8.1	Location of Van Mon commune	173
8.2	Aluminium production chain	177
8.3	Diseases or health problems suffered by sampled households in 2006	186
9.1	The Asian highlands comprising the Tibetan Plateau, mountain ranges, the main rivers and river basins	199
10.1	Map of Mekong uplands and Montane Mainland Southeast Asia	218
10.2	The risk, hazard, exposure, vulnerability relationship	223
10.3	Knowledge, risk and consent	224

TABLES

3.1	Resettlement progress in Son La and Lai Chau provinces as of May 2008	50
3.2	Communication strategies and outcomes in Son La	54
4.1	Strategies employed by the Hmong	85
5.1	Areas of different land according to the watershed classification scheme of Ministry of Science and Technology in 1996 (adopted by the RFD)	94
6.1	Core studies on gender in aquaculture and fisheries from the Mekong region	118

6.2	Impacts of commercialization and capture-to-culture transitions on gender relations	135
8.1	Outputs of aluminium production in Man Xa	175
8.2	Household borrowing in the surveyed year	179
8.3	Net cash income sources of different social groups of households/year/capita in 2006	179
8.4	Selected heavy metal concentration in rice paddy and irrigation canals	183
8.5	Man Xa villagers' visits to Van Mon Health Clinic	185
8.6	Diseases and health problems in Man Xa 2004–2005	187
8.7	Number of cancer-related deaths	187
9.1	Possible effects of climate change on ecosystems and society from mountain top to delta in the Mekong region	207
10.1	The risks–rights nexus during flash flood management cycle	232

Boxes

1.1	The Mekong region and this book	2
3.1	Types of resettlement in the Son La programme	50
4.1	Creating cultural spaces within watersheds	74
4.2	The *ntoo xeeb* ceremony in Ban Phui Nua	79
5.1	Watershed classification of the Karen communities in the Upper Mae Hae watershed	97
5.2	Special projects in the uplands	99
5.3	Activities of the MHNC	104
5.4	Local government in Thailand	105
10.1	The Mekong uplands, peoples and livelihoods	221

Contributors

Robert Arthur is a senior consultant at MRAG Ltd, a UK-based consultancy company. His research concentrates on fisheries as complex social-ecological systems and the interfaces between research and development, natural and social sciences and policy and action, with a focus on the politics of natural resource management and approaches such as adaptive management and co-management. He has been working in the Mekong region since 1999.

Nathan Badenoch has been working on natural resource governance in the uplands of mainland Southeast Asia through a range of academic and policy research organizations since 1998. His interests include social networks, community-based resource management and indigenous knowledge, in addition to transboundary environmental governance. Badenoch obtained his PhD in Southeast Asian Area Studies from Kyoto University's Graduate School of Asian and African Area Studies. He is currently an Associate Professor affiliated with the Hakubi Project at Kyoto University, based at the Center for Southeast Asian Studies.

Geeta Bhatrai Bastakoti has a Masters degree in Gender and Development Studies from the Asian Institute of Technology, Bangkok, Thailand. She has been involved in a gender and water insecurities project while working with the Unit for Social and Environmental Research, Chiang Mai University, Thailand. Currently she is the team leader for a research project focusing on gender issues in small-scale aquaculture in Nepal.

David Blake has 12 years of experience working in Thailand and Lao PDR in a variety of natural resource management and rural development fields, including both state and non-state organizations. He is presently enrolled as a PhD candidate at the School of International Development, University of East Anglia, UK, while researching cross-scalar drivers of irrigation development in northeast Thailand.

Chanagun Chitmanat is an assistant professor at the Faculty of Fisheries Technology and Aquatic Resources, Maejo University, Chiang Mai, Thailand. His research focuses on aquaculture, fish diseases, aquatic environment and aquaculture extension.

Rajesh Daniel has been working and writing on environmental issues in Thailand and the Mekong region for over 15 years. He is a researcher and editor with M-POWER and the Unit for Social and Environmental Research (USER), Faculty of Social Science, Chiang Mai University, Thailand. His research focuses on land and forest governance in the Mekong region with special emphasis on upper tributary watersheds, indigenous peoples and local knowledge systems.

Nga Dao is co-founder and director of the Center for Water Resources Conservation and Development (WARECOD), an environmental organization in Vietnam, and is working towards her PhD in human geography at York University in Canada. Her focus is on the political ecology of water governance and agrarian transformation in uplands Vietnam and mainland Southeast Asia.

Mark Dubois is a Research Fellow at the WorldFish Center, Mekong Office, Cambodia. He has been working in the Mekong region since 1999 on water resources and fisheries management and governance issues. His current research interests include social enquiry, power relations and participatory mechanisms for integrating knowledge with action.

Philippe Floch is an associated researcher with the Institute of Hydraulics and Rural Water Management at the University of Natural Resources and Applied Life Sciences, Vienna, Austria, and has a doctorate in Land and Water Resources Management and Engineering. He has worked for six years in Southeast Asia, focusing on water policy design and implementation, irrigation management and river basin planning.

Richard Friend is an independent consultant based in Bangkok, Thailand, and is the research leader of the fisheries and livelihoods research theme of M-POWER. He has a background in social anthropology and development studies and has been working in the Mekong region for over 15 years on issues related to livelihoods, governance and policy analysis, with an interest in water resource management and fisheries and, more recently, climate change.

Santita Ganjanapan teaches at the Department of Geography and Regional Centre for Social Science and Sustainable Development, Faculty of Social Sciences, Chiang Mai University. Her academic interests are sustainable development, community-based natural resource management, environmentalism, gender and environment and agro-food systems.

Le Thi Van Hue is a researcher/lecturer at the Center for Natural Resources and Environmental Studies, Vietnam National University, Hanoi, Vietnam. Her research focuses on natural resource management, land tenure and gender. She has been involved in projects that aim to reach the poorest people, especially women and ethnic minorities, and to contribute to their socio-economic development. Her work has touched on issues relating to the demographic, ethnic, social, socio-economic and cultural practices prevalent within craft villages, forestland and relevant policy and legislation in Vietnam.

Kate Lazarus has worked in the Mekong Region for 10 years on water resources management and environmental governance issues. She leads the M-POWER Dialogues Working Group and is coordinating the development of multi-stakeholder platforms for the Challenge Program on Water and Food. Her research and policy work focuses on the role of China in the Mekong region, environmental flows, nexus of rights and environment, stakeholder engagement and hydropower governance. She currently lives in Lao PDR.

Louis Lebel is the founding Director of the Unit for Social and Environmental Research (USER), Faculty of Social Science, Chiang Mai University, Thailand. He has broad research interests including ecology, epidemiology, political science and governance. He regularly contributes to the international global change science programmes. He has lived and worked in Thailand for most of the past 18 years.

Phimphakan Lebel is a researcher at the Unit for Social and Environmental Research (USER), Faculty of Social Science, Chiang Mai University, Thailand. Her research has focused on shrimp and fish aquaculture commodity chains including studies of gender relations in natural resource management in Thailand.

Prasit Leepreecha has a PhD in Anthropology from the University of Washington, Seattle. Presently, he is a researcher at the Center for Ethnic Studies and Development, Faculty of Social Science, Chiang Mai University. His main interests include ethnic minorities in northern Thailand and mainland Southeast Asia, identity, culture change, tourism, nationalism and ethnic minorities, globalization and ethnic minorities.

Songphonsak Ratanawilailak is an M-POWER research fellow hosted by the Unit for Social and Environmental Research (USER), Faculty of Social Science, Chiang Mai University, Thailand. He is engaged in research on upland watershed governance focusing on ethnicity, local institutions and dialogue in resource management in northern Thailand.

Bernadette P. Resurreccion is an Associate Professor in Gender & Development Studies at the Asian Institute of Technology (AIT) in Thailand and teaches and carries out research on livelihoods, migration and natural resource management. Her research on these themes has focused in countries such as the Philippines, Thailand, Vietnam and Cambodia and she has recently published a co-edited book *Gender and Natural Resource Management: Livelihoods, Mobility and Interventions* (Earthscan and IDRC, 2008).

Edsel E. Sajor is an Associate Professor at the School of Environment, Resources and Development at the Asian Institute of Technology (AIT) in Bangkok, Thailand. He has a PhD and MA in Development Studies from the Institute of Social Studies, The Hague, in The Netherlands. He teaches governance, environmental conflict resolution and social research methods. His areas of research and academic publication are on urban-rural linkages, peri-urban studies, land and water development and urban environmental management issues. His geographic study focus is Southeast Asia, especially the Mekong Region countries.

Mith Somountha is an official at the Ministry of Rural Development in Thailand with an MA in Sustainable Development from Chiang Mai University, Chiang Mai, Thailand. Her research has focused on hydropower dams, fisheries and livelihoods, including gender studies in natural resources management.

Tran Tri Ngoc Trinh is a researcher at the Soil Science and Land Management Department, College of Agriculture and Applied Biology, Can Tho University, Vietnam. Trinh's research has focused on the application of GIS (geographic information systems) and Remote Sensing in Land Evaluation and Planning in Vietnam.

Tran Van Ha is a senior researcher at the Institute of Anthropology, Vietnam Academy of Social Sciences (VASS). He has a BA in Anthropology from Hanoi University (1977) and a PhD (1996) from the Institute of Anthropology. Tran is a member of the Vietnam Association of Anthropology, has written widely on ethnic minorities of Vietnam and is an active member of the Vietnam Rivers Network (VRN). He is also the editor of the book *Indigenous Knowledge Research in two Fishing Villages along the Red River* (Mai Hoang Publisher, 2008).

Jianchu Xu is a senior scientist at the World Agroforestry Center (ICRAF), China and a professor in the Kunming Institute of Botany affiliated with the Chinese Academy of Sciences (KIB/CAS). His work focuses on indigenous people, biodiversity and cultural survival, forest management, community livelihoods, climate change and watershed governance.

Preface: About M-POWER

This book is a product of M-POWER, which stands for the Mekong Program on Water, Environment and Resilience. The network brings together people committed to improving local, national and regional governance in Cambodia, China, Laos, Burma/Myanmar, Thailand and Vietnam.

The ultimate goals of M-POWER are improved livelihood security and human and ecosystem health in the Mekong Region. We contribute to this by focusing on improving and democratizing water governance.

Our action research, practical policy support and facilitation involve pursuing fair and effective governance, which takes account of possible rewards, voluntary and involuntary risks, and rights and responsibilities of all authorities and stakeholders. We are committed to ensuring that water-related negotiations and decision-making are more fully informed and transparent.

M-POWER is primarily supported by the efforts and resources of the partner organizations who choose to cooperate in this transnational effort to improve governance. Substantial financial support for 2006–2010 has been given by the Consultative Group on International Agricultural Research (CGIAR) Challenge Program on Water and Food (CPWF) via resources from Echel-Eau (Government of France) and the International Fund for Agricultural Development (IFAD). For more information, see www.mpowernet.org.

This is the third of a four-volume multi-authored book series on water governance in the Mekong Region produced by M-POWER. The first volume, *Democratizing Water Governance in the Mekong Region*, was published in 2007 by Mekong Press, Chiang Mai, Thailand. The second volume, *Contested Waterscapes in the Mekong Region: Hydropower, Livelihoods and Governance*, was published in 2009 by Earthscan, UK and USA. The series provides critical perspectives and addresses water governance issues of contemporary relevance including contested discourses, policy directions, alternative development scenarios and action-research agendas in sharing, developing and managing water resources. The first volume provided a baseline, state of knowledge review of the politics and discourses on water governance; volume two focused on hydropower, livelihoods and governance; this third volume is about social justice and rights with case studies on watershed politics, fisheries and hydropower. The series editors are Rajesh Daniel and Louis Lebel.

Acknowledgements

This volume has been conceptualized, shaped and written by partners in the Mekong Program on Water, Environment and Resilience (M-POWER) water governance network, with authors from across the Mekong region. The editors acknowledge and thank all the writers for their efforts and also the many people who provided constructive feedback on drafts.

We are grateful to the many individuals who contributed their time and effort to reviewing and providing comments on individual chapters. We especially wish to thank Eric Baran, Chris Barlow, David Dore, John Dore, Rebecca Elmhirst, Barry Flaming, Richard Friend, Kim Geheb, Dipak Gyawali, Peter Hansen, Erik Harms, Kanokwan Manorom, Carl Middleton, Francois Molle, Le Hung Nam, Andrew Noble, Jeff Rutherford and Chusak Wittayapak. We are also especially grateful for the support of Louis Lebel, programme leader at M-POWER and Director of the Unit for Social and Environmental Research at Chiang Mai University.

Special thanks go to Rajesh Daniel at the Unit for Social and Environmental Research (USER) for assistance with publishing this volume and David Lau for help with copyediting. Many thanks to Tim Hardwick, Claire Lamont and Hamish Ironside at Earthscan for their support and patience.

Acronyms and Abbreviations

ADB	Asian Development Bank
AR4	IPCC's Fourth Asssessment Report
ASEAN	Association of Southeast Asian Nations
BDP	MRC's Basin Development Programme
CPWF	Challenge Program on Water and Food
DOSTE	Department of Science and Technology (Vietnam)
DWR	Department of Water Resources (Thailand)
EGAT	Electricity Generating Authority of Thailand
EIA	environmental impact assessment
ENSO	El Niño/Southern Oscillation
EVN	Electricity of Vietnam
FDI	foreign direct investment
FPIC	free, prior and informed consent
GCM	general circulation model
GDP	gross domestic product
GHG	greenhouse gases
GIS	geographic information systems
GLOF	glacial lake outburst floods
GMS	Greater Mekong Sub-region
GWP	Global Water Partnership
ha	hectare
IBFM	integrated basin flow management
ICHRP	International Council on Human Rights Policy
ICRAF	International Centre for Research on Agroforestry (World Agroforestry Centre)
IFAD	International Fund for Agricultural Development
IFI	international financial institution
IUCN	International Union for Conservation of Nature
IWMI	International Water Management Institute
IWRM	integrated water resources management
kg	kilogram

LDOF	landslide dam outburst flood
LEP	Law on Environmental Protection (Vietnam)
masl	metres above sea level
MDB	multilateral development bank
MDG	Millenium Development Goal
MHNC	Mae Hae Network Committee
MMSEA	Montane Mainland Southeast Asia
MONRE	Ministry of Natural Resources and Environment (Vietnam)
M-POWER	Mekong Program on Water, Environment and Resilience
MRC	Mekong River Commission
MRC-TAB	MRC Technical Advisory Body for Fisheries Management
MRCS	MRC Secretariat
MSN-WU	Mae Sa Nga Watershed Unit
MSP	multi-stakeholder platform
MW	megawatt
MWRAS	Mekong Water Resources Assistance Strategy
NESDB	National Economic and Social Development Board
NGO	nongovernmental organization
NGPES	National Growth and Poverty Eradication Strategy
NIAPP	National Institute for Agriculture Planning and Projection (Vietnam)
NSEDP	National Socio Economic Development Plan
PADEK	Partnership for Development in Kamphuchea
PM	prime minister
PVC	polyvinyl chloride
RBC	River Basin Committee
RFD	Royal Forestry Department (Thailand)
RID	Royal Irrigation Department (Thailand)
RP	Royal Project
SEA	Strategic Environmental Assessment
TAI	The Access Initiative
TAO	Tambon Administration Organization
TAO-AC	TAO-Administrative Committee
TB	tuberculosis
TEI	Thailand Environment Institute
UN	United Nations
UNDP	United Nations Development Programme
UNDRIP	UN Declaration on the Rights of Indigenous Peoples
UNECE	United Nations Economic Commission for Europe
UNEP	United Nations Environment Programme
UN-REDD	United Nations – Reduced Emissions from Deforestation and Forest Degradation
USAID	United States Agency for International Development

USER	Unit for Social and Environmental Research
VCG	Vietnam Consultative Group
VND	Vietnamese dong
VRN	Vietnam Rivers Network
VUSTA	Vietnam Union of Science and Technology Associations
WARECOD	Center for Water Resources Conservation and Development (Vietnam)
WB	World Bank
WCC	Watershed Classification Committee
WFP	World Food Programme
WWF	World Wildlife Fund

1

Water Governance and Water Rights in the Mekong Region

Nathan Badenoch, Kate Lazarus,
Bernadette P. Resurreccion and Nga Dao

The Mekong region has come to represent many of the important water governance challenges faced more broadly by the mainland Southeast Asia region. The development and environment issues raised in this complex arena of economic growth and integration highlight the importance of water, at all levels: locally, nationally and regionally. While technology and engineering dominated the period of rapid growth in the 1980s and 1990s, moving into the 21st century there has been increased awareness among researchers of the urgent need to refocus thinking on the processes of decision-making through which the region's waters are utilized, developed, transported, degraded or conserved (Lebel et al, 2007b; Molle et al, 2009). Despite macroeconomic figures suggesting that the region is well on the way to poverty eradication, a disaggregated look at socio-economic development indicators tells a story of inequality across sectors of society (ADB, 2007a,b; UNDP, 2009; CIE, 2010). The need for access to clean water at predictable times is common to all in society, but access to the resource is predicated on access to the processes of governance that determine who gets water, in what quality and quantity, and at what time. Demands on the water resources have grown (ADB, 2007a,b), creating new situations of scarcity that are dependent not only on topography and climate, but also the capacity of societies to manage claims on the flows of water. Research has shown that when water crises occur, it is most frequently a result of continual mismanagement and insufficient governance (Biswas, 2010).

This book is concerned with the governance of water resources in the Mekong region, which is considered in its broadest sense here to include the countries of Burma/Myanmar, Cambodia, China, Laos, Thailand and Vietnam (see Box 1.1).

> ## Box 1.1 The Mekong region and this book
>
> The Mekong region covers the territories, economies, politics and peoples of Burma/Myanmar, Cambodia, China, Laos, Thailand and Vietnam and is home to over 325 million people. Major rivers include the Irrawaddy, the Nu-Salween, the Chao Phraya, the Lancang-Mekong and the Red River. These rivers provide lifelines to numerous peoples whose language, religion and culture have been heavily shaped by the dynamic nature of the region.
> The region is home to a range of natural habitats. Upper catchments provide livelihood opportunities for small-scale agriculture. The rich biodiversity within the region, especially the inland fisheries in the Mekong River, is fundamental to the viability of natural resource-based rural livelihoods of the people. These livelihoods are founded on the integrated use of a wide range of natural resources that have adapted to the seasonal changes of the region. However, knowledge of the region's rich human and natural diversity is still considered rudimentary as scientists continue to discover new languages and animal species (WWF, 2008). Moreover, the Mekong region is slated to be significantly impacted by climate change in the future (WWF, 2009; TKK and SEA START RC, 2009). As the economy and people are intrinsically linked to the natural resource base, the consequences of climate change add pressure to the impacts also expected from infrastructure development projects that will have huge impacts on the people of the region in the short to medium term (Keskinen et al, 2010).

Taking a wide regional frame of reference makes sense from the social, economic and hydrological perspectives, as it recognizes the inter-linkages between people and nature beyond the watershed level. But some of the steepest challenges arise from the politics of transboundary cooperation, coordination, decision-making and problem solving. There is a range of institutions involved in trying to address regional governance in the Mekong, all with differing mandates (see IUCN et al, 2007; Osborne, 2009; Lee and Scurrah, 2009). For example, there is continued debate over the effectiveness and relevance of the Mekong River Commission (MRC), a regional body with a mandate over the 'lower' Mekong part of the region, but lacking the membership of China (IUCN et al, 2007; Dore and Lazarus, 2009; Lee and Scurrah, 2009), while national governments push on with the support of projects and programmes bilaterally and through the Asian Development Bank's Greater Mekong Subregion Initiative that directly or indirectly affect the region's water resources and those who depend on them.

Despite discussion of the social and environmental impacts of large-scale water infrastructure projects, development of water resources maintains a central place in the economic vision of the region's national-level decision-makers. But concerns over the social, environmental and economic outcomes of interventions in the water sector, bolstered by an increasingly high profile debate about water for livelihoods, persist among researchers, activists and many groups who might potentially be affected. In taking up 'social justice and water rights' at a regional level, this book seeks to explore different aspects of the relationship among people and the processes of governance. Water governance is more than just about water and related services but involves issues of access to decision-making and information, participation and justice, and ethnicity and gender. Equally important are the linkages between the processes of governance, the outcomes of the decisions taken and the possibilities that are created at different levels and scales by actors in specific contexts.

The chapters of this volume draw their analyses and perspectives from a broader range of water governance challenges faced by the region: large-scale infrastructure projects to impound or move large quantities of water, impacts of market transition on agriculture and aquaculture, competition between industrial and domestic water uses, and the vast uncertainties associated with climate change. These analyses are indicative of a number of the most important challenges facing a region where national economies show dynamism but the political status quo remains firm. From this point of view, one practical question is whether the comparatively conservative governance characterized by the nation states of the Mekong region is able to recognize the challenges, define the agenda and make the necessary adaptations to meet the needs of an economically dynamic but volatile regional economy. The allocation and enforcement of rights among sectors of society forms a foundation for the establishment of predictable and equitable social relations, especially with regards to claims on resources. If nested within robust governance processes, rights can provide a framework for resilience in the face of uncertainties that accompany socio-economic development.

WATER AND RIGHTS

In development discourse, there is a tension between 'water as a right' and 'water as an economic good'. The question of water rights has held a central position within efforts to reform the legal and institutional frameworks around the globe (Bruns and Meinzen-Dick, 2005). Despite this global trend, the Mekong region as a whole has not made a decisive move towards clear legal frameworks that define rights over access to and management of water. Rights in this sense are construed as legally recognized and backed claims over a resource.

Water rights are established and maintained through processes of negotiation. Bruns and Meinzen-Dick (2001) offer four areas in which rights play a central role in the negotiation of water resources: 1) renegotiating rights, which involve intervention and reorganization of regimes; 2) formalization of rights, strengthening existing claims and regimes; 3) basin governance, comprising the negotiation of users linked in natural systems; and 4) inter-sectoral transfers, which entail negotiation across systems. Indeed, the area of water rights is one in which pluralistic legal frameworks are essential to capturing the complex dynamics of interaction among competing claims (Neef et al, 2006).

One neoliberal approach to addressing the problems of water allocation in an age of growing competition has been to design water-pricing schemes. This essentially entails promoting the commodification of a resource that in many settings was previously treated as a common property resource not managed by market-based mechanisms. The economic principle of allocating water according to capacity to pay assumes much about the social aspects of resource management. Observing the adverse impacts that this allocational principle had on the rural poor and other marginalized sectors of society, a global response within civil society to the perceived problems of this type of neoliberal intervention in the water sector led to a rush to a 'water as a human right' banner. While legal rights assuring access to water are surely an important part of any effort to reform water governance, critics argued that asserting a human right at the global or regional level did very little to improve the situation of access to water or the decisions that affect water. An analytically more nuanced and strategically more sound approach, it was argued, would be a shift back to a framework that treats water not as an inherent right of humanity and not a commodity, but as a common-property resource (Bakker, 2007).

The authors in this book take a broad approach to water rights, writing about not only rights directly associated with access to water but including other rights that affect people's ability to access the areas of governance, through formal and informal means, that affect water resources decision-making. It is also necessary to examine how rights are created, negotiated, asserted, contested, ignored or denied at various levels of decision-making, especially for those that are adversely affected. An improved understanding of these key dynamics should be clearly linked to the social equity and justice outcomes that are observed. This means moving beyond the legalistic exercise of creating laws and decrees, and moving to an analysis that is more firmly rooted in real-life, real-time challenges of implementing, adapting and revising these arrangements for water rights, among the sectors of society that face the most serious barriers to exercising those rights. This perspective underscores the importance of outcomes in terms of equity rather than efficiency.

Equity and Social Justice

With equity at the forefront of our analytical focus, we are working with a theoretical concept of social justice, in a region with vastly differing economic development trajectories, diverse political traditions and administrative systems, and complex mosaics of human and biological diversity. Social justice is not concerned with merely a narrow conception of the benefits to individuals, but rather with what is good for the society as a whole. This requires an inclusive approach that examines a broad range of actors and interests (Capeheart and Milovanovich, 2007). Nonetheless, in a region where state-driven models of resource-based development have been dominant, we also see a need to give special focus to those sectors of society that are marginalized from the critical areas of governance. In his recent critique of the mainstream philosophy of justice expounded by Rawls (1971), Sen has argued for a more realistic 'idea of justice' – one that focuses on eliminating injustices, to replace past conceptions that were concerned more with finding an ideal form of justice. Although the authors of this volume do not seek to propose any models of social justice for the region, Sen's (2009, p106) assertion that 'a theory of justice must have something to say about the choices that are actually on offer, and not just keep us engrossed in an imagined and implausible world of unbeatable magnificence', offers a challenge to unpack the prescriptive and theoretical frames of governance principles such as participation and access, and move beyond legalistic approaches to mediating not only the interactions between people and their environments, but between people who share a resource as well.

Two main aspects of justice (Capeheart and Milovanovic, 2007) are helpful here: distributive and retributive justice. The two are clearly linked and necessary for a comprehensive conception of social justice, but separation of the two assists us in identifying the entry points of discussing justice in concrete and contextualized terms. Distributive justice is 'how rewards and burdens are distributed', that is to say an interactive set of decisions about what is fair and what is not, in effect the processes of negotiating values by which outcomes are assessed. Retributive justice is the 'mechanisms of accountability', meaning the interactions that take place within a set field of values to ensure that fairness is maintained. The idea of distributive justice is considered a dynamic, process-oriented framework that allows us to explore a number of different cases. Thus the practical concern with social justice discussed by this book's authors seeks to uncover some of the 'choices on offer', examining them within the socio-economic, cultural and political realities of the Mekong region.

WATER GOVERNANCE AND WATER RIGHTS

'Water governance' is the arena in which these issues are played out. The Global Water Partnership definition of water governance, which has been widely cited, is the range of political, social, economic and administrative systems that are in place to develop and manage water resources, and the delivery of water services, at different levels of society (Global Water Partnership, 2002). We also refer to water governance as:

> *the ways in which society shares power with respect to decisions about how water resources are to be developed and used, and the distribution of benefits and involuntary risks from doing so. It includes the full spectrum of influences from shaping agendas and deliberating options through the design of institutions and laws through the way these are implemented in the practices of day-to-day management of water.*
> (Lebel et al, 2010)

Water governance has been widely studied by academics and development researchers, with particular emphasis on the role that institutions play, who has access and who participates in decision-making realms (e.g. Biswas et al, 2005; Lebel et al, 2005; Molle and Wester, 2009; and Dore and Lebel, 2010).

Yet, in reality, modes of water governance are highly contested across sectors of society (Molle, 2008). Managing water resources for the socio-economic well-being of people and the robust functioning of ecosystems is a complex challenge. Stakeholders are diverse and have competing demands for water, as well as being highly differentiated in terms of influence over decision-making. Insufficient frameworks for information provision at the national and regional level limit the scope for inclusive governance processes surrounding water decision-making.

Furthermore, significant imbalances in economic and political power colour the reality of water decision-making. Scale and level politics play a role in defining who participates and with what role in deliberations around water decision-making (Dore and Lebel, 2010). The end result is that trade-offs are not well calculated, risk is distributed to sectors of society that cannot bear the consequences (Lebel and Bach, 2009) and the long-term viability of socio-economic development founded on the region's water resources is threatened. Inevitably, how these complex interactions play out is a product of the governance structures and processes through which decisions over water management and development are made.

In particular, spaces for participation by local people are not neutral but are shaped by power relations that both surround and enter them (Cornwall, 2004). Building on the concept of negotiation spaces, there is a need to look closely at levels of power in terms of how spaces are created, the levels of engagement within these spaces and the degree in which power relations play a role within them (Hickey and Mohan, 2004). Participation spaces for poor people are constantly opening and closing, thus creating a dynamic relationship between and among stakeholders as one asserts and determines their human right to manage and access water. However, in practice, the emphasis on the economics imposed by national development plans means that the principle of efficiency dominates and the principle of equity falls victim to the forces of the market. Concerns of this type were addressed in the UNDP 2006 Human Development Report that looked 'beyond scarcity', and highlighted the poverty–power nexus in decision-making over water resources.

It is commonly argued that the role of water rights in contributing to basic livelihood security and social justice is of eminent importance (Bessette, 2006). This book takes livelihood security as a departure point for its exploration of justice and rights, given the prominent position of poverty alleviation within the socio-economic development plans and programmes implemented by national government with the support of international development agencies, bilateral development assistance, and non-government organizations. Many of the decisions taken at this level have direct and indirect impacts on people's access to water. There are great imbalances of power between the mechanisms of the state and the voices of the citizens embodied in the decision-making structures, governance processes and flows of resources involved in national development efforts. However, it is important to keep in mind that non-state actors are diverse within countries and across the region, especially as regards those sectors of society commonly referred to as 'local people' or 'the poor'. For example, urban Thai women in the Bangkok metropolis will have very different interests and resources than Hmong women living in the mountains near the Lao–Vietnam border. Similarly, Mekong delta rice farmers plot their livelihood strategies within a completely different framework from upland farmers in northern Vietnam whose livelihoods are based on cash crops and collection of non-timber forest products. Despite these and a myriad of other social differences, the common thread lies in the relative difficulty they have in influencing the big decisions that may affect their access to water and other livelihood resources. The architecture of water governance is seriously skewed towards state-centric models of decision-making, at both the national and regional levels. Thus, while recognizing these constraints to inclusive governance, it is necessary to keep observations grounded in the realities of each setting.

Similarly, those actors that hold a large section of the decision-making authority are often not united. While we often speak of 'national governments' as if they were monolithic, one of the most common critiques of water resources development projects is the difficulty of coordination between relevant agencies. This includes not only administrative and bureaucratic hurdles, but also involves agency interests and objectives that are at odds with each other.

The international financial institutions (IFIs), such as the World Bank and the Asian Development Bank, which drive many of the development agenda items with conditional funding, are also at the mercy of national governments to implement projects and achieve their own development goals. Thus, governments and IFIs find themselves criss-crossing through areas of shared interest and divergent objectives with their national government partners. Water governance is a particularly good example of this dynamic, where IFIs must make an effort to implement internationally recognized principles of good practice (including transparency, consultation and participation). However, in recent years 'new' financiers, mainly from Asian countries (such as China, Thailand, Vietnam, Russia, Korea, Malaysia and India) have begun to dominate investments in the water sector (Rutherford et al, 2008; MRC, 2010). The host governments often criticize the international standards pushed by IFIs as being onerous and cumbersome while civil society groups target these IFIs for poor practices. The new financiers are not yet bound by such standards in their overseas activities, even though they may be required to follow them in their home countries. They are seen as bringing a different kind of investment package to the table: one that does not have benchmarks of compliance with human rights, democratic ideals and environmental protection regulations, but is built on relationships and friendship (Rutherford et al, 2008).

Securing the rights of people with claims over water resources requires governance structures that are inclusive and not only create and allocate rights to different sectors of society, but, more importantly, protect them in the face of competing interests. Indeed, the main challenge to water governance is in creating processes that facilitate the exercise of rights through equitable negotiation and decision-making. Thus in order to speak of rights concerning water, there is need for a broader treatment of the bundles of rights that are needed to ensure access to decision-making and negotiation that affect water and peoples' access to it (Bruns et al, 2005).

Decisions concerning hydropower, irrigation, domestic water supply, inland and coastal fisheries and flood management contribute to the social, economic and environmental transformations of the region. Despite their livelihoods being highly dependent on the region's water and other resources, there continues to be a large and growing segment of poor, landless farmers, small fishers, and migratory labourers, many of whom are women and minority ethnic groups, struggling to support themselves and their families. They struggle not only to get rights of access but also find themselves bearing the brunt of the risks and vulnerabilities from these water governance policies and decisions.

This book focuses on the complex nature of water rights and social justice in the Mekong region. The chapters delve into the diverse social backgrounds that frame the various realities of water governance in the region, in the hope of bringing to the forefront some of the local nuances required in the formulation of a larger vision of justice in water governance. It is hoped that this contextualized analysis will deepen our understanding of the potential of, and constraints on, water rights in the region, particularly in relation to a Mekong-specific articulation of social justice.

STRUCTURE OF THE BOOK

The chapters in this book place particular emphasis on the complex and multiple human-induced developments that are increasingly contributing to widening gaps in wealth, resource access and power. This volume is divided into four parts that follow this introductory chapter. The first part focuses on **participation in decision-making**, particularly the lack of participation by locally affected people in infrastructure development projects such as hydropower or diversion schemes and their socio-economic consequences. The second part looks at issues of **social differences and access**, the relationships around the use and management of water resources, particularly in terms of gender and ethnicity in the creation of institutions to assert rights at the local level. The third part discusses the negotiation of **competing demands and protecting the rights of the marginalized** over water between different sectors such as hydropower and fisheries, industry and transboundary issues. The final part reflects on **climate change and the rights of the vulnerable**, focusing on poor people in both uplands and lowlands for developing regional policies to address adaptation to climate change in the region.

Part I: Participation in decision-making

Participation can have different meanings for different people. There are different levels of participation. Development decisions are often taken without the meaningful participation of poor people and the most vulnerable (women and ethnic groups) are the most affected by the impacts of such decisions. What does it take to successfully engage people around development projects in the Mekong region? Why are some groups of people being left out of decision-making processes? Institutionalizing meaningful deliberation and dialogue between a range of actors such as governments, developers, regulators, planners, businesses and civil society on focused issues is needed in order to bring all sides of an issue to discussion. Chapters 2 and 3 specifically address the lack of participation by poorly represented actors in water resources decision-making process, its socio-economic consequences. For example, it is common practice in the Mekong region for decisions to be made around large-infrastructure projects without local community knowledge or input into the process.

In Chapter 2, Philippe Floch and David Blake explore the multifaceted issues associated with participation of local people in decision-making around real and complex water management situations. They draw from a situation that has been brought to discussion numerous times in Thailand – the Lao–Thai Water Transfer project. They assert that without the participation of those that are directly impacted by development schemes (such as hydropower or water transfer schemes) decisions cannot be made effectively and thus do not incorporate the needs and wishes of the people (mostly local people). The authors reflect on the issues of good governance and the role in which participation in planning processes can contribute. This project aims to transfer a massive amount of water from 'water rich' Laos to 'water scarce' northeast Thailand. They assert that the participatory reality has not been matched with the adopted rhetoric espoused by major actors in water resources policy-making in Thailand.

The role of participation in resettlement policies in Vietnam is discussed by Tran Van Ha in Chapter 3, based on analysis of the Son La Hydropower Project, which upon its completion will be the largest dam in Southeast Asia. Tran explores the notion that information and community dialogue are important steps in providing affected people with information so that they can readily participate in important processes that contribute to gaining public acceptance. Informing people is seen as the first step and, importantly, ensuring information in local languages is essential. The Son La example provides a useful window on the current directions in participation policy and practice in Vietnam, because it is setting a precedent for the way in which the government of Vietnam handles resettlement programmes throughout the country. It is expected that the decisions made around this project will continue to have long-term implications for water governance in Vietnam. Moreover, Tran locates his analysis within the historical context of dams in Vietnam in order to give a more nuanced understanding of how the Son La experience should be understood. Comparing various aspects of participation processes with earlier dam construction experience, he is able to provide a more dynamic view of changes in the practices of water governance and signal some of the key aspects of social justice that should be considered in Vietnam.

Part II: Social differences and access

Despite decades of debate and intervention to improve the legal and institutional frameworks for participation in development, access to some of the arenas of decision-making is still not a reality for many groups. Local people continue to lose their rights to access, use and management of water resources that they depend upon for farming, fishing, drinking and sanitation. Furthermore, many local people are faced with challenges and limited abilities to claim or defend their rights or speak out against development projects that impact their livelihoods. Society positions them in ways that evoke differences and cleavages in class, gender, ethnicity and age, and which are reproduced and sustained by socio-cultural discourses, norms and practices that translate into hierarchies and forms of inequity in their access to resources, including water. Three chapters in this volume explore this domain of the role of gender and ethnicity in water governance, drawing on the broad range of experience with community involvement in water and watershed resources in northern Thailand.

Chapter 4 by Nathan Badenoch and Prasit Leepreecha explores the responses of the Hmong people, an important ethnic group in the Mekong region that has been at the centre of agricultural and economic transition in upland landscapes, particularly in northern Thailand. The chapter discusses the efforts of the Hmong to establish themselves as legitimate and responsible managers of upland watershed resources, while at the same time securing access to important land and water resources. The evolution of irrigation institutions to manage conflict among upstream water users during the dry season has led to the development of a more privatized regime of access rights that are negotiated with other users in an area where no legal rights over water govern the competing claims. At the same time, the Hmong have given new meaning to cultural symbols and used these in highly visible events to demonstrate an 'indigenous' tradition for responsible watershed management. Thus there are simultaneous efforts to produce workable rights regimes at the local level, with commitments to responsibility at a broader level of watershed discourse. Interestingly, while the media have reported enthusiastically on conflict between the Hmong and lowland Thai farmers, the Hmong have chosen to assert themselves through this dual approach mixing rights and responsibilities not directly with local Thai farmers, but rather within the upland areas themselves and at a more symbolic level of public relations.

Rajesh Daniel and Songphonsak Ratanawilailak, in Chapter 5, discuss the politics of upland watershed management showing how upland farmers in northern Thailand, predominantly ethnic communities, are coping not only with resource scarcity such as seasonal water shortages but also with state conservation laws and official development strategies. Cases from the upper watershed areas of Mae Hae and Khun Kan illustrate the negotiation and contestation between different definitions of 'watershed' and those who use it. The cases offer examples of on-the-ground efforts of local level actors including individual leadership, watershed networks that cut across administrative boundaries and local administrative organizations. The cases explore how these individuals, networks and administrations shape their definitions and perspectives, and further their diverse objectives, in upland watershed management.

Gender, like ethnicity, is a critical dimension of social difference that alongside class often has important consequences for access to water and the services water flows and aquatic ecosystems provide. Louis Lebel and colleagues in Chapter 6 seek to understand how gender relations affect, and are affected by, commercialization and transitions from capture fisheries to aquaculture in the Mekong region. Some actors believe that the emergence of an aquaculture industry is part of the solution to addressing the claims over the poor condition of, and weak prospects for capture fisheries. Others see it as part of the problem, in which commercialization leads to environmental degradation. This polarization is an oversimplification of the range of social-ecological relationships present. Inclusion of aquaculture in livelihoods may be highly complementary to existing capture fisheries or it may be in direct competition for natural resources and labour. Either way, capture fisheries and aquaculture have distinct features that impact on divisions of labour and responsibilities between women and men. The benefits accrued to women from their engagement in fisheries and aquaculture activities, however, are less predictable as they are influenced by specific cultural, market and other factors. Four factors appear to be important to understanding whether and how engagement leads to material benefits or empowerment: novelty, diversity, mobility and proximity. The implication for social justice is that engagement in commercial aquaculture or fisheries is not automatically an empowerment pathway or a burden-multiplier.

Part III: Competing demands and protecting the rights of the marginalized

Regional, national and local development priorities, not to mention conservation concerns, clash at multiple scales across the region. Water resources are at the centre of economies across these scales and decisions over how water should be allocated is fundamentally fraught with competition among diverse users. The outcomes of resource competition are a factor of economic and political power, negotiation processes and other social and cultural dynamics. The Mekong region is faced with competing demands over water use – water for irrigation, water for infrastructure, water for energy, domestic use and ecosystem service provision, to name just a few. The two chapters in this section discuss the trade-offs associated with competition at the local and regional levels of water resources management.

In Chapter 7, Robert Arthur, Richard Friend and Mark Dubois further illuminate the current debates in the Mekong region over hydropower versus fisheries. As hydropower developments are accelerated, particularly along the Mekong mainstream, debates are looming over how to address the loss of abundant fisheries that are so important to local livelihoods. Recent reports (e.g. Friend et al, 2009; Dugan et al, 2010; ICEM, 2010) indicate that the fisheries cannot be substantially replaced by aquaculture and if lost, pose the problem of food insecurity and the loss of important nutritional sources for millions of people. Arthur and colleagues discuss just what is at stake and what might be lost by considering how fisheries contribute to development that meets the needs of the people of the Mekong. They emphasise the importance of food sovereignty as a local issue in the discourse about trade-offs in water decision-making. In particular the authors explore experiences from Lao PDR, a country with a rich capture fishery, but also endemic food crisis, and a national policy commitment to both poverty reduction and significant hydropower development.

Le Thi Van Hue and Edsel Sajor discuss in Chapter 8 the tensions between private sector development and environment and livelihoods in Vietnam as the country rapidly industrializes and grows economically. Here, a craft village involved in household-based artisanal production is used as a case study to discuss the right of people to livelihood pursuits and their notion of natural water bodies as common goods with regards to industrial water use and wastewater management. They emphasise the need for mechanisms that enable community and households to engage with public authorities in collaborative and participatory planning and negotiate decision-making to handle the tensions between household livelihood interests, community welfare and ecological sustainability and macro-economic goals pursued by the government of Vietnam.

Part IV: Climate change and the rights of the vulnerable

Climate change is upon us. In the last five years the Mekong region governments have produced numerous climate-related documents including action plans, adaptation plans and strategies, attended hundreds of workshops and conferences, and received millions of dollars to address the looming climate change 'issue'. But still the upland and lowland people of the region continue to be in vulnerable situations as the region experiences intense weather conditions ranging from typhoons and floods to droughts and they risk losing their livelihoods and social structures. Two chapters explore the uncertainty of climate change and implications for access to water, local rights and social equity at the regional level.

Jianchu Xu in Chapter 9 examines climate change in the Asian highlands, which, combined with growing societal impacts, looks set to pose far-reaching socio-economic implications for the diverse ecosystems and societies in the Mekong region. In addition to the numerous natural risks they face, people living in fragile ecosystems or areas vulnerable to the effects of extreme weather events have been placed in the socio-political position of being guardians of their upland areas and therefore responsible for many of the environmental impacts that are perceived by lowland, mainstream society to stem from changes in upper watershed areas. The chapter argues the need for more inclusive governance processes that on the one hand facilitate access to the necessary preventative and adaptive capacities needed in all sectors of society, and on the other hand guarantee the rights of all those at risk to influence these processes, regardless of location within the socio-political and ecological landscapes.

Chapter 10 by Jianchu Xu and Rajesh Daniel provides a conceptual study that can bring together at least three threads that in the authors' view are often found lacking in the climate change discourse; i.e. the uplands, poor and marginalized people and their disproportionate risks, and the human rights dimension. The chapter uses a rights-based framework to try to address climate-related vulnerabilities including both natural- and human-induced hazards for the upland peoples, especially the poorest. They pay particular attention to climate rights, broadening the discussion from specific, visible impacts on human rights to those caused by global change over longer periods of time and with more holistic appraisals.

Conclusion

Given the central importance of water to livelihoods, economies and ecological systems at all scales, the outcomes of water governance are a crucial concern for justice within society. Improving the institutions for governance of water resources at the regional, national and local levels must seek more than mere efficiency and effectiveness. Decisions over how water is used should contribute to advancing the goal of water justice. But understanding of rights and justice in Mekong region water governance remains largely articulated in terms of broad normative principles that are divorced from the diverse and dynamic realities of the region. This book is an attempt to provide some context to the thinking of justice and rights in water governance in the Mekong region. In exploring some of the local contexts in which water governance plays out, the collection of chapters demonstrates that moving beyond a simple conception of social justice as the logical outcome of a general idea of water rights can help move towards a more practical vision of change in the governance of the Mekong region's water resources.

References

ADB (2007a) *Asian Water Development Outlook 2007: Achieving Water Security for Asia*, Asian Development Bank, Manila

ADB (2007b) 'Inequality in Asia', Asian Development Bank, Manila

Bakker, K. (2007) 'The "Commons" Versus the "Commodity": Alter-globalization, anti-privatization and the human right to water in the Global South', *Antipode*, vol 39, no 3, pp430–455

Bessette, G. (2006) *People, Land and Water*, Participatory Development Communication for Natural Resource Management, IDRC, Ottawa, Canada

Biswas, A.K. (2010) 'Water: Crisis due to scarcity or poor governance?'. Closing Keynote Address, Global Water Conferences, Frankfurt, Germany, 20 May

Biswas, A.K., Varis, O. and Tortajada, C. (2005) *Integrated Water Resources Management in South and South-East Asia*, Oxford University Press, New Delhi

Bruns, B.R. and Meinzen-Dick, R. (2001) 'Water rights and legal pluralism: Four contexts for negotiation', *Natural Resources Forum*, vol 25, pp1–10

Bruns, B.R. and Meinzen-Dick, R. (2005) 'Frameworks for water rights: An overview of institutional options', in B.R. Bruns, C. Ringler and R. Meinzen-Dick (eds) *Water Rights Reform: Lessons for Institutional Design*, International Food Policy Research Institution, Washington DC

Bruns, B.R., Ringler, C. and Meinzen-Dick, R. (eds) (2005) *Water Rights Reform: Lessons for Institutional Design*, International Food Policy Research Institution, Washington DC

Capeheart, L. and Milovanovic, D. (2007) Social justice: Theories, issues and movements, Rutgers University Press, New Brunswick, NJ

CIE (2010) 'Economic benefits of trade facilitation in the Greater Mekong Subregion', Center for International Economics, prepared for Australian Agency for International Development (AusAID), Canberra and Sydney

Cornwall, A. (2004) 'Spaces for transformation? Reflections on issues of power and difference in participation in development', in S. Hickey and G. Mohan (eds) *Participation: From Tyranny to Transformation?*, Zed Books, New York

Dore, J. and Lebel, L. (2010) 'Deliberation and scale in Mekong Region water governance', *Environmental Management*, vol 46, no 1, pp60–80

Dore, J. and Lazarus, K. (2009) 'De-Marginalizing the Mekong River Commission', in F. Molle, T. Foran and M. Käkönen (eds) *Contested Waterscapes in the Mekong Region: Hydropower, Livelihoods and Governance*, Earthscan, London

Dugan, P.J., Barlow, C., Agostinho, A.A., Baran, E., Cada, G.F., Chen, D., Cowx, I.G., Ferguson, J.W., Jutagate, T., Mallen-Cooper, M., Marmulla, G., Nestler, J., Petrere, M., Welcomme, R.L. and Winemiller, K.O. (2010) 'Fish migration, dams, and loss of ecosystem services in the Mekong Basin', *Ambio*, vol 39, pp344–348

Friend, R.M., Arthur, R.I. and Keskinen, M. (2009) 'Songs of the doomed: The continuing neglect of capture fisheries in hydropower development in the Mekong', in F. Molle, T. Foran and M. Käkönen (eds) *Contested Waterscapes in the Mekong Region: Hydropower, Livelihoods and Governance*, Earthscan, London

Global Water Partnership (2002) 'Introducing Effective Water Governance', mimeo.

Hickey, S. and Mohan, G. (2004) 'Towards participation as transformation: Critical themes and challenges', in S. Hickey and G. Mohan (eds) *Participation: From Tyranny to Transformation?*, Zed Books, New York

ICEM (2010) 'MRC SEA for hydropower on the Mekong mainstream: Impacts assessment (opportunities and risks)', discussion draft report by the International Centre for Environmental Management, p249

IUCN, TEI, IWMI and M-POWER (2007) 'Exploring water futures together: Mekong region waters dialogue', Report from regional dialogue, Vientiane, Laos

Keskinen, M., Chinvanno, S., Kummu, M., Nuorteva, P., Snidvongs, A., Varis, O. and Västilä, L. (2010) 'Climate change and water resources in the Lower Mekong River Basin: Putting adaptation into the context', *Journal of Water and Climate Change*, IWA Publishing, vol 1, no 2, pp103–117

Lebel, L. and Bach, T.S. (2009) 'Risk reduction or redistribution? Flood management in the Mekong region', *Asian Journal of Environment and Disaster Management*, vol 1, no 1, pp23–39

Lebel, L., Garden, P. and Imamura, M. (2005) 'Politics of scale, position and place in the governance of water resources in the Mekong Region', *Ecology and Society*, vol 10, p18

Lebel, L., Dore, J., Daniel, R. and Koma, Y.S. (eds) (2007) *Democratizing Water Governance in the Mekong Region*, Mekong Press, Chiang Mai

Lebel, L., Bastakoti, R.C. and Daniel, R. (eds) (2010) CPWF Project Report. Enhancing Multi-Scale Mekong Water Governance. Project Number PN 50, Consultative Group on International Agricultural Research (CGIAR) Challenge Program on Water and Food, Chiang Mai

Lee, G. and Scurrah, N. (2009) 'Power and responsibility: The Mekong River Commission and lower Mekong mainstream dams', A joint report of the Australian Mekong Resource Centre, University of Sydney and Oxfam Australia

Molle, F. (2008) 'Nirvana concepts, narratives and policy models: Insight from the water sector', *Water Alternatives*, vol 1, no 1, pp131–156

Molle, F. and Wester, P. (2009) 'River basin trajectories: An inquiry into changing landscapes', in F. Molle and P. Wester (eds) *River Basin Trajectories: Societies, Environments and Development*, Comprehensive Assessment of Water Management and Agriculture, Central for Agricultural Bioscience International (CABI), Wallingford, UK

Molle, F., Foran, T. and Käkönen, M. (eds) (2009) *Contested Waterscapes in the Mekong Region: Hydropower, Livelihoods, and Governance*, Earthscan, London

MRC (2010) 'The MRC and Hydropower', available at www.mrcmekong.org (last accessed 25 October 2010)

Neef, A., Chamsai, L. and Sangkapitux, C. (2006) 'Water, tenure in highland watersheds of northern Thailand: Managing legal pluralism and stakeholder complexity', in Lebel et al (eds) *Institutional Dynamics and Stasis: How Crises Alter the Way Common Pool Resources are Perceived Used and Governed*, RCSD/Chiang Mai University, Chiang Mai

Osborne, M. (2009) 'The Mekong: River under threat', Lowy Institute for International Policy, Lowy Institute Paper 27, New South Wales

Rawls, J. (1971) *A Theory of Justice*, Harvard University Press, Cambridge, MA

Rutherford, J., Lazarus, K. and Kelley, S. (2008) 'Rethinking investments in natural resources: China's emerging role in the Mekong Region', Heinrich Bolle Foundation (HBF), WWF, International Institute for Sustainable Development (IISD)

Sen, A. (2009) *The Idea of Justice*, Penguin, London

TKK & SEA START RC (2009) 'Water and climate change in the Lower Mekong Basin: Diagnosis and recommendations for adaptation', Water and Development Research Group, Helsinki University of Technology (TKK) and Southeast Asia START Regional Center (SEA START RC), Chulalongkorn University, Water and Development Publications, Helsinki University of Technology, Espoo, Finland

UNDP (2009) 'Thailand human development report 2009', UNDP, Bangkok

WWF (2008) 'Greater Mekong: Close encounters – New species discoveries 2008', WWF Greater Mekong Programme, Laos

WWF (2009) 'The Greater Mekong and climate change: Biodiversity, ecosystem services and development at risk', WWF, Bangkok

Part I

Participation in Decision-making

2

Water Transfer Planning in Northeast Thailand: Rhetoric and Practice

Philippe Floch and David Blake

INTRODUCTION

Over the last decade, calls for good governance and open (and more democratic) planning processes have increasingly started to infiltrate the water sector, with the need for public participation in project selection, design and operation appearing prominently in virtually all studies, policy recommendations and scholarly papers (e.g. Rogers and Hall, 2003). This is consistent with the observation that 'the quality of governance pervades public decision-making relating to policy formulation, resources allocation, legislation, rule enforcement and adjudication, making it the most important single influence on the shape and pace of institutional change in the water sector' (Svendsen et al, 2005). Also, for the last 20 years or so, contemporary analysis of planning has focused on the role of actors, their interaction and patterns of communication, as well as the distribution of power and agency within society, thereby challenging more traditional (but ironically, strongly modernist) approaches based on ideas of techno-centric and rational-comprehensive planning. Concepts like accountability, transparency and legitimacy are now being talked about in the international water resources governance discourse with regularity, and the ascendancy of the theme of good governance has brought politics into the mainstream water resources development discourse 'through the backdoor' (Mollinga, 2008).

In this chapter we reflect on the ways planning – occurring within a particular social, economic, cultural and political context – can play out with regard to both the theoretical backdrop against which it is designed, and the competing narratives adopted by actors in the water sector of Thailand. Our focus here is primarily on irrigation development, which constitutes the main focus of several ambitious state-led water projects targeted at the northeast region of Thailand, with potentially profound socio-environmental impacts both on the immediate region and, on a wider scale, the larger Mekong River Basin of which it is part.

We do so by first reflecting on integrated water resources management (IWRM) and collaborative planning in general, taking the illustrative case of Thailand. Based on these entry points, we look at a recent water resources development to divert water from the Nam Ngum river basin in Laos into Thailand. Our focus examines the rhetoric adopted by state and non-state actors engaged in planning and nascent stakeholder participation processes. In the last sub-section of this chapter, we reflect on the gap between theory and practice of IWRM practices in general, and the notion of collaborative planning in particular.

IWRM AND COLLABORATIVE PLANNING: FRAMING THE DISCUSSION

Within the water sector, the discourses concerning participation in planning and management of water resources are firmly embedded in the wider arguments for IWRM. According to one often-cited definition (GWP, 2000) IWRM is 'a process which promotes the coordinated development and management of water, land and related resources, in order to maximize the resultant economic and social welfare in an equitable manner without compromising the sustainability of vital ecosystems'. More recently USAID (2007) defined IWRM as 'a participatory planning and implementation process, based on sound science that brings in stakeholders together to determine how to meet society's long-term needs for water and coastal resources while maintaining essential ecological services and economic benefits'. This latter definition establishes a clear link between the need for participatory approaches in planning and implementation of hydraulic infrastructure, while also stressing the need for broad involvement of stakeholders, which is uniformly recognized as a precondition for sustainable development. Some argue that 'IWRM cannot be achieved without public participation' (Özerol and Newig, 2008), while others counter this view by pointing out that 'participatory processes have been increasingly approached as technical, management solutions to what are basically political issues' (Gujit and Shah, 1998) and talk about a 'new tyranny' of participation (Cooke and Kothari, 2001).

Planners, donors and governments quickly made use of the catchphrases resulting from an increased literature on IWRM processes and water governance, adopting a more participatory language in recommendations and planning documents, and calls for good governance feature in most (if not every) policy recommendation and planning document. For example, the Asian Development Bank (ADB) (2003) promoted 'improved governance' (among other key requirements for an integrated approach to water resources management); The Hague Ministerial Declaration (2000) called for 'governing water wisely to ensure good governance, so that the involvement of the public and the interest of all stakeholders are included in the management of water resources', and in Thailand the current 10th Economic and Social Development Plan (2007–2011) specifies that 'the country should have quality environment and sustainable natural resources management under the good governance principle' (NESDB et al, 2008).

Many of the changes that have taken place in the dominant paradigms to managing natural resources in general, and water more particularly, have been influenced by changes in planning theory. Natural resource planning has long been based upon the rational-comprehensive model: scientifically based and expert-driven, this model implicitly assumed consensus on a single objective, availability of all the data needed to support decisions and seemingly unlimited financial resources and time (Lachapelle et al, 2003). This approach has increasingly been questioned and successively altered and replaced, as the credibility of the rational planning model was widely debated and challenged by scholars of planning theory (e.g. Lim, 1986; Lachapelle et al, 2003). Increasingly, planning theorists have started to debate methods and programmes to encompass issues of discourse and inclusiveness (Fainstein, 2000).

As early as the 1980s, Forester (1980) showed that relations of power are part of all planning exercises and argued that these power relations manifest themselves in 'unnecessary distorted communication'. While the modernist planning project allowed planners to provide 'facts', the more inclusive and communicative forms of planning now called on planners to work with contending parties and their competing claims (Saarikoski, 2002). Since then, scholars have constantly refined understandings of planning as a communicative enterprise. As Huxley (2000) summarized:

> *the communicative planning literature rejects as unrealistic the idea of planning as technical and apolitical, and, indeed, technical and political neutrality are seen to be incapable of achieving planning's reformative goals. Instead, planners and planning systems need to be responsive to differences, to be genuinely participatory, and to strive to create deliberative contexts, that, as far as possible, minimize inequalities of power and knowledge.*

At the same time that theory started to embrace participatory or collaborative forms of planning, calls for more inclusive and democratic practice started to permeate water policy documents. It has been observed that within the Mekong region there is a general call for participation, community/collaborative models of governance and multi-stakeholder platforms or approaches focusing on negotiations (e.g. Dore, 2007; Molle, 2007). One clear example of the close link between the newly advocated modes of water resources planning was provided by the United Nations Development Programme (UNDP, 2006) which argued that 'multi-stakeholder engagement processes' provide a key mechanism for avoiding political conflict and that 'one multi-stakeholder approach to water management is Integrated River Basin Management'.[1]

IWRM AND THE PLANNING RHETORIC IN THAILAND

As a member of the United Nations, the government of Thailand has ratified the Johannesburg Plan for Implementation at the World Summit for Sustainable Development in 2002; by 2006 the national policy and planning developments of Thailand's National Economic and Social Development Board had been revised to be in agreement with the United Nation's Framework on Sustainable Development, including the Ninth (2000–2006) and 10th National Economic and Social Development Plans (2007–2011); and both Agenda 21 and Local Agenda 21 were implemented between 1997 and 2006. All of these developments prominently featured participatory approaches to planning, development and management of natural resources.

These wider changes are reflected in Thailand's approaches to managing water resources, with IWRM being the officially sanctioned water management paradigm of the country. The Eighth National Economic and Social Development Plan (1997–2001) aimed 'to promote effective management, involving the collaboration of various different sectors of society, so as to achieve greater balance in ecosystem and environments' and elaborated that 'opportunities will be provided for local people and organizations to play a greater role in natural resource and environmental conservation' (NESDB, 1996). The National Water Vision Statement for Thailand, released in 2001, concurred with the need for more participatory approaches in water management, and contained all parts of a progressive approach to water resources management. It reads, 'by the year 2025, Thailand will have sufficient water of good quality for all users through efficient management and an organizational and legal system that will ensure equitable and sustainable use of water resources, with due consideration given to the quality of life and the participation of stakeholders' (Ti and Facon, 2001). The 9th Economic and Social Development Plan argued that 'the government will try to set up the institutional framework of water administration with users' participation by transforming its strategy and operating style in order to give opportunity to stakeholders, especially local people, to participate in water resources management' (Sethaputra et al, 2001). The current 10th National Economic and Social Development Plan for the years 2007–2011 adheres to this participatory rhetoric and targets the building of 'strong communities with an inherent strategy to improve communities through increased participation, planning and knowledge management' (ITD, 2007). Equally the National Sustainable Development Strategy (NESDB et al, 2008) calls for 'developing models and replicating all sectors' integrated participatory water source management and rehabilitation' and more generally aims 'at developing Thailand to be a participatory society in development based on honesty, transparency and impartiality'.

All of the above is evidence of attempts by policy-makers and water management specialists to foster IWRM approaches – as best practice – in Thailand, with a participatory rhetoric adopted in virtually all documents, guidelines and public speeches by a host of actors both within the country's administration and beyond. The ADB, for example, observed that 'Thailand's rich historical relationship with water has evolved in recent years into a dynamic programme of integrated water resources management with participation of local stakeholders' and that 'Thailand has established itself as a leader in pioneering a participatory approach to water resources management in river basins' (ADB, 2004). Also, the quasi-non-governmental Thai Water Resources Association (with strong links to the Department of Water Resources) stated that 'the integrated water resources management principle has been incorporated into the water resources management process of Thailand' (ESCAP, 2005).

Participation is a very broad term capturing many meanings and interpretations (Heyd and Neef, 2004), stretching from passive participation to information sharing, to consultation and institutionalized participation. Collaboration, as advocated by planning scholars, needs to differ from passive participation (with people being told what is happening), participation in information giving (with people participating by answering questions) and participation by consultation with people being consulted, while external agents define both the problems and solutions. Participatory approaches to planning require that problems are jointly defined, that solutions are broadly discussed and assessed and that different types of knowledge are factored into the process of planning. Neef (2009) also argued that the concept of polycentric governance 'provides a useful tool to understand many of the transformation processes within water governance regimes' by changing the responsibilities and capabilities of state and non-state actors, while shifting power and resources amongst these actors.

THE LAO–THAI WATER TRANSFER

Setting the scene

Planners concerned with agricultural development in northeast Thailand have long been interested in the possibilities of water augmentation to a region that has been consistently portrayed as water scarce since the end of World War II (Molle and Floch, 2008; Blake and Floch, 2009). This, perhaps, explains the tendency for politicians and state planners to be concerned with promoting developments to supplement existing supplies with water from outside the region ever since the early development phases of the region's drainage basins.[2] Initially these water imports were designed to be sourced from the ambitious and massive Pa Mong 'multi-purpose' dam[3] on the mainstream Mekong River. However, as geopolitical, socio-economic and environmental concerns slowly derailed that dam project, planners continued to work on other, no less ambitious, plans to transfer water to Thailand's northeast region. Briefly, these post-Pa Mong projects included:

- the 'Green Isaan' project in the 1980s (Molle et al, 2009);
- the 'Khong-Chi-Mun Irrigation Project' in the 1990s (Sneddon, 2003); and
- the 'Water Grid' project in 2003 (Molle and Floch, 2008).

This brief chronology not only sets the background for this chapter, it also frames the historical setting that a planning team 'scoping for options in joint Lao-Thai water management' entered. It is this last initiative that this chapter is mostly concerned with, partly due to its relevance to current Thai national water resources planning options and discourse.

Visions to divert water from the lower sections of the Nam Ngum River in Lao PDR to Thailand date back to at least 1994 when Sanyu Consulting proposed to divert water from the lower section of the Nam Ngum River to northeast Thailand (Southeast Asia Rivers Network, 2002), diverting water both to the Lam Pao Reservoir and the Nam Songkhram Basin. This initiative followed the early phases of the implementation of the Khong-Chi-Mun irrigation project, with a cascade of in-stream weirs along the main rivers of northeast Thailand implemented to capture run-off for agricultural production through large-scale pumping schemes. Importantly, while the implementation of the Khong-Chi-Mun project not only triggered popular protest within Thailand, it also was part of the underlying reason the Interim Mekong Committee (the predecessor of today's Mekong River Commission) was discontinued (Molle and Floch, 2008). In essence, with the former basin arrangements vesting veto rights with member countries to challenge mainstream abstractions of water resources (planned under the Khong-Chi-Mun project), Thai water bureaucrats argued that a transfer from the Nam Ngum to northeast Thailand would be considered merely a tributary development, to be treated solely under bilateral negotiations. Sanyu Consultants Inc. (2004) later presented another version of this development option, which extended the 1994 study by including possible diversions from the other tributaries in the Laos, including the Xe Bang Fai, all of which later became prominent parts of the Water Grid proposal. Significantly, it was the 2004 version of the water transfer plan that was re-discussed under the Mekong Water Resources Assistance Strategy (MWRAS) (described below).

Mekong Water Resources Assistance Strategy (MWRAS)

In October 2004, the World Bank (WB) and the ADB announced that they would undertake a comprehensive assessment of the possible long-term sustainable use of water resources in the Mekong River Basin, as a basis for preparing a Mekong Water Resources Assistance Strategy (MWRAS) (BIC, 2005). MWRAS aimed to provide guidelines on the management and utilization of water resources in the Mekong Basin, 'ensuring that the principles of "balanced development" are incorporated into the water resources projects' (AMRC, 2007). In its inception phase, MWRAS focused on three target areas suitable for testing the proposed development activities, including the border section of the Mekong Basin between Laos and Thailand. Seven criteria justified the selection of these target regions (WB and ADB, 2006), including:

1   economical and financial attractiveness;
2   potential to deliver multiple benefits while protecting key social and environmental values;
3   easy identification of trade-offs;
4   creating an environment to develop stronger governance institutions;
5   aggregate financial capabilities;
6   potential to build regional trust; and
7   being broadly endorsed by all stakeholders, non-governmental organizations (NGOs) and civil society.

In the same month that an MWRAS working paper was published, a consultant to the WB also prepared an inception report titled 'Scoping the options for joint water resources development and management between Lao PDR and Thailand in the Mekong Basin' (Consultant Team, 2006). As the study detailed with regard to the overall MWRAS strategy 'work in these three regions[4], taken together, would stimulate a growing sense of cooperation, and that each country can receive benefits in a win-win perception because it generates economies of scale, builds regional trust, delivers multiple benefits and helps set up and strengthen governance institutions'. Also, the report explained that 'these three regions have been brought forward by the countries, and also through the bottom-up planning process of the Basin Development Program (BDP) of the Mekong River Commission Secretariat (MRCS)'. Indeed, the MWRAS working paper summarized that 'the Royal Thai Government has identified better water provision as its second highest national priority', especially in northeast Thailand.

Planning for water diversions into northeast Thailand had gained considerable momentum prior to the MWRAS. In 2003, the Royal Thai Government under then-Prime Minister Thaksin Shinawatra, announced that the country was to engage in an ambitious megaproject to increase the area under irrigation in the country from 4.7 million ha by a further 16.5 million ha within five years, to enable farmers to cultivate crops and access water around the year. The largest intended beneficiary of that project was the people of northeast Thailand, the electoral stronghold of Thaksin's Thai Rak Thai party. In early 2004, however, the project came under increasing criticism (Molle and Floch, 2008): academics questioned its economic profitability while social and environmental activists predicted waterlogging, salinity and social equity problems would arise. Further, the massive amount of capital needed in realizing this vision triggered bureaucratic rivalries, as both the Royal Irrigation Department (under the Ministry of Agriculture and Cooperatives) and the Department of Water Resources (under the newly created Ministry of Natural Resources and Environment) competed to oversee the project with separate plans produced by each department (Samabuddhi, 2004). When in September 2006 a military coup ended the supremacy of the Thaksin government, the Water Grid plans were brought to a temporary halt. Significantly, it is within this highly politicized planning environment situated between a surprise military coup and bureaucratic infighting that the appointed consultants engaged in studies to promote greater regionalization in water sharing futures.

The rhetoric

When the MWRAS consultancy team commenced work in 2006, it issued an inception report stating that the study would follow 'a holistic, consultative and inclusive "sub-basin" approach with a regional perspective and involve interactive phased processes' (Consultant Team, 2006). In more detail, the study team explained that the approach would: 1) take into consideration not only technical/ engineering possibilities but socio-economic and environmental dimensions; 2) reflect views of wider stakeholders through consultation and public meetings; and 3) pay attention to upstream developments and possible impacts on the downstream basin. As such, the wording of the inception report borrowed from best practice of both contemporary planning theory (consultation, inclusive, interactive, public meetings) and the wider IWRM paradigms (multi-disciplinary, upstream-downstream interactions). But the most critical aspect of the planning endeavour was the attempt to forward a notion of collaborative and participatory planning, including wider civil society in an apparent effort to adhere to a best-practice planning routine.

The justifications forwarded in a second working paper by the consultancy team (Consultant Team, 2007a) summarized the rationale as a textbook case for large-scale water transfer:

> *The rich volume of water is left untapped in Laos while most of the tributary basins on the other bank of the Mekong, Northeast Thailand, suffer from water shortage every year during the dry season. Both Laos, which is water rich, and Northeast Thailand, which is water-stressed, could realize benefits equally by the consorted [sic] efforts of formulation and implementation of joint development and management of water resources and water related activities in the tributaries of and the mainstream of the Mekong.*

The rhetorical case for a transboundary diversion, thereby, was inscribed and set initially: the 'water-scarce' and suffering northeast Thailand was to receive water from 'water rich' Laos, thereby generating a supposed win-win situation. However, the benefits for Laos were not readily apparent and were mostly seen in terms of potential payments of resources royalties and options to foster foreign direct investment from neighbouring Thailand (including potentials for large-scale concession to the private sector for developing irrigated agriculture, such as biofuel production). Equally in Thailand, benefits from such ambitious investments had earlier been contested when the Water Grid was initially proposed under the Thaksin administration, and did not feature prominently in the consultant's report. While the study highlighted other possible options for joint water development,[5] it is the water transfer (maybe more than any other component) that highlights the challenges of implementing IWRM principles within a historically politicized planning environment. Of course, such a large-scale undertaking would require a considerable amount of additional infrastructure development to transfer, distribute and make use of water resources (see Figure 2.1).

The study team argued that the positive impacts of this development option would include both flood mitigation in the Huay Luang and Huay Suai basins (relatively small Mekong tributaries), as well as 'assured dry season water supply for all residents for all purposes; especially for villages and farm households in Chi and Mun Basins' (Consultant Team, 2007a). In essence, this mirrored the political announcements made earlier by representatives of the Thai Government advocating for the Water Grid project (Molle and Floch, 2008). However, associated social and environmental costs were not discussed, even at the most basic level of estimates.[6] This neglect is, perhaps, explained by a prior study on the Nam Ngum Diversion (Sanyu Consultants Inc., 2004) that estimated total infrastructure costs at US$660 million, or around 0.5 baht per cubic metre of transferred water, excluding potential resource royalties to the Government of Laos. Simultaneously with the Department of Water Resource's planned cross-border water transfer project, RID have been compiling plans of their own to tap the waters of the Mekong bordering Loei province and divert water via a series of reservoirs, tunnels, canals and other infrastructure into the Chi-Mun and Nam Songkhram Basins (Thanopanuwat, 2008). Thanopanuwat (2008) stated that, '…the use of Mekong water will definitely be a powerful strategy in fighting against Northeast Thailand's serious water shortage problem. It will serve the National Interest in every aspect' [sic].

Public scrutiny and stakeholder involvement

In February 2007, the MWRAS team of consultants invited selected stakeholders to a 'provincial workshop' at the Charoen Thani Princess Hotel in Khon Kaen, northeast Thailand. The audience was composed of a restricted mix of government officials from various ministries and departments, some representatives from the nascent regional River Basin Committees (RBC),[7] Thai academics, and a handful

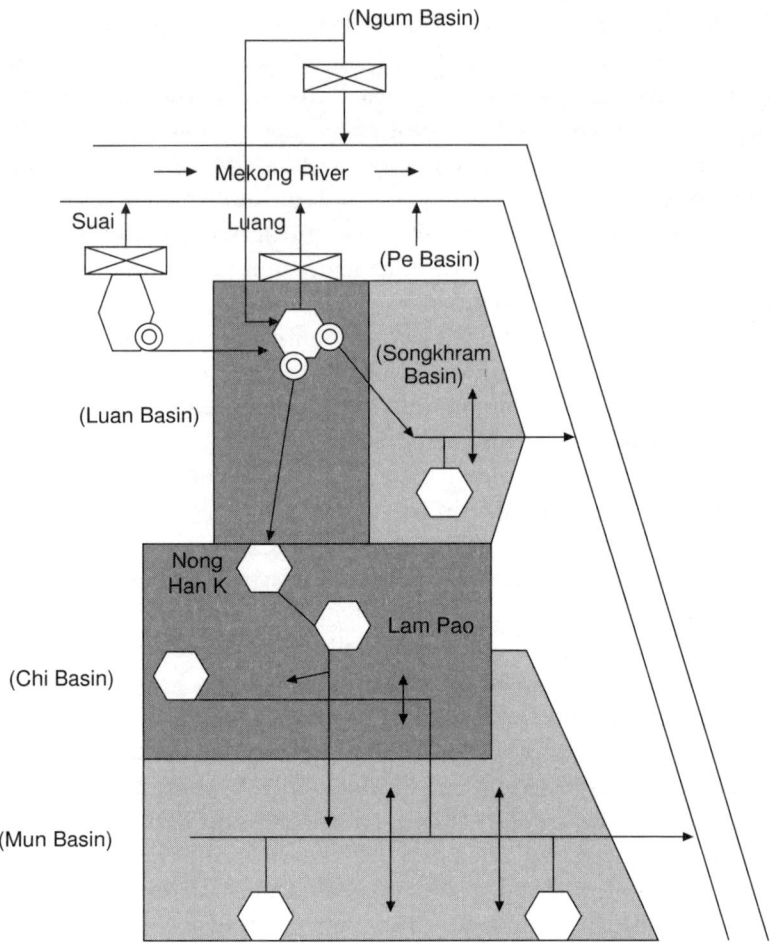

Figure 2.1 *Option for Nam Ngum–Northeast Thailand water transfer*

Source: Consultant Team, 2007a

of NGO and civil society representatives (most notably from the International Union for Conservation of Nature and the Thailand Environment Institute), with the consultant team[8] aiming to air their study recommendations against a 'critical' audience.

As might have been expected, the criticism that emerged within the discussion was manifold. A representative of the Khong[9] RBC argued that the study should have been transparent in the first place, and that he failed to see any valuable benefits arising from the project. Another representative from the same river basin committee felt that only benefits were presented, and questioned how many people would be negatively impacted. Others asked the consultants to provide evidence of successful irrigation projects in northeast Thailand. A representative of the Chi RBC indicated that a number of prior studies had already been conducted on this issue that were not factored in the consultants' report and that earlier large-scale irrigation projects had not contributed to poverty alleviation. Representatives of the International Union for Conservation of Nature (IUCN) pointed to possible impacts on regional wetlands-based livelihoods and other natural resources while also pointing to previous negative experiences arising from the semi-completed Khong-Chi-Mun project. One underlying question during the meeting was summarized by a member of the Khong RBC: 'Did the study ask the right questions? Do the [northeast] people really need water more than other things?' (Consultant Team, 2007b). However, while the responses of the invited participants were perhaps predictable within the Thai context of contested water resource projects, the consultation proceedings were interrupted by a group of NGO protesters who took the microphone, demanding their voices be heard, arguing that, 'people be informed about the project' and to 'cancel the water transfer between Thailand and Laos', maintaining that 'Thai farmers are getting poorer as a result of more big projects' (Protest Communiqué, 2007). It was apparent that the 'public' stakeholder consultation had excluded sections of the northeast Thai non-state actors with definite interests in water resources planning and decision-making processes. The views of Laotian stakeholders were not considered either, nor was there recognition of the fact that such a public protest would be virtually untenable in Laos.

In the aftermath of the Khon Kaen stakeholder workshop the momentum of this particular episode in planning the Nam Ngum water diversion project was lost, while the attention of the project portfolio resulting from the MWRAS shifted towards other potential support within the Mekong basin. At the same time, and within Thailand, the diversion of water from the Mekong or from neighbouring Laos continued to rank high on the agenda. In June 2008, the short lived Prime Minister Samak (widely seen as a proxy of deposed PM Thaksin Shinawatra) announced his government's intention to invest in megaprojects, including water diversions (Ekachai, 2008). More recently, the project to siphon water from the Nam Ngum River won cabinet approval (Wipatayotin, 2008), although it is far from clear to what extent the Laos authorities are on board with the plans. A senior Laos irrigation official who delivered a paper at a Food and Agriculture Organization regional irrigation conference in Vietnam suggested that it may not be necessary or desirable to transfer water to Thailand out of tributaries, but instead sell a reserved share of its flow input to the Mekong to its neighbour, claiming this was a 'holistic view' and in line with MRC's multilateral agreement on use of Mekong flows (Pheddara, 2007). Lately, under the present Democrat Party led government, both the Royal Irrigation Department (RID) and the Department of Water Resources (DWR) are engaged in preparing new detailed proposals to import water either directly from Laos or by drawing from potential increased dry season flows within the Mekong mainstream itself that are expected to result from hydropower generation upstream in Yunnan. However, it would seem the Thai and Laos riparian populations and civil society at large are not being informed about the justifications for, and details of, these plans or the chance to engage in meaningful dialogue about regional water futures.

WATER RESOURCES PLANNING: BETWEEN THEORY AND PRACTICE

While it goes well beyond the scope of this chapter to scrutinize the plethora of claims associated with water development plans of the complexity of such inter-basin or cross-border water diversions, it is equally important to discuss how planning played out with regard to the ideal set out in policy documents and the rhetoric of consultants' reports.

Collaborative and participatory planning embraces collective decision-making in ways that enhances more transparent and accountable forms of governance, allowing all participants to debate and interrogate others (Brand and Gaffkin, 2007). Literature suggests that planners can foster distortion-free communication and that such communication can result in a consensus based agreement. This has been questioned by scholars, and Mouffe (1999) even argued that free and unconstrained public deliberation of all matters of common concern is conceptually impossible. As Brand and Gaffkin (2007) explained 'power differentials, a reality well recognized by many advocates of collaborative planning, cannot be dissolved through logical argumentation'. To be fair, this should not be mistaken with the fact that the pursuit of informed public deliberation could contribute to better decision-making in water resources planning. Yet, better planning relies mostly on resolving power inequalities, an issue that is hardly addressed in planning guides and policy documents as this would explicitly challenge current elites and entrenched power structures.

Thailand's recent water development history has been marked by events of public deliberation that turned out as mere public announcement forums (Sneddon, 2003; Molle and Floch, 2008; Chang Noi, 2009), aiming to 'educate' an 'uninformed' or 'uneducated' rural population about the merits of a particular investment. This is at odds with attempts to build what Novotny (1999) calls 'socially robust knowledge', or public participation that goes beyond state actors (and consultants engaged by them) defining both problems and solutions. Without specific recognition of power differentials in society, IWRM and its variants such as integrated basin flow management (IBFM) are in danger of becoming no more than a managerial exercise, similar to some of the blueprint participatory processes in rural development (Neef, 2009) or be reduced to a 'Nirvanic' type concept (see Molle, 2008).

The brief workshop in Khon Kaen outlined above, which was admittedly only to present intermediate project findings, provides an illustrative case in point.

Trying to channel 'participation' through the institutional framework of the MRC (and particularly its National Mekong Committees) ensured that the starting hypothesis (the problem definition) was in agreement with national goals guiding water policy in Northeast Thailand. This bias was augmented by the very definition (MRC, 2005) of stakeholders: internal stakeholders that are the governing bodies of the MRC and the principle line agencies of each member country, and external stakeholders that constitute non-state bodies such as NGOs, implementing partners, civil society organizations, policy advocators, research groups, individual media and other groups who have interest or stakes to gain or lose (Middleton, 2007). As the invitation list confirms, the consultants aimed to open up discussion to internal stakeholders, inviting members of concerned line agencies, the Thai National Mekong Committee, River Basin Committees (with a mix of state and non-state local actors) and only a few additional resource person invited (including the authors of this chapter). A broad range of external stakeholders, representing the extremely active wider Thai NGO community, were not invited. However, by opening up the discussion (both voluntarily and involuntarily), the planning team found itself confronted in a value discussion that they had aimed to avoid.

It was argued that MWRAS 'turned a blind eye to more complex issues' including barriers to participation and decentralization of power, vested interests and competition within and between ministries and regional politics (Middleton, 2007). This is evident in the Inception Report prepared for the joint water resources development that aimed to de-politicize the social history of water transfer planning in northeast Thailand. This meant that at the time the consultant team offered its findings, 'options' were only marginally part of the efforts, but discussions quickly surrounded the history and record of water resources development in the region, which had either been ignored or removed from sight.

Added to this is the recognition that parties in a dispute usually not only disagree over single knowledge claims, but also employ essentially different frameworks through which they select evidence and provide it with meaning and interpretation (Saarikoski, 2002). Within the available and highly heterogeneous existing knowledge base concerned with irrigation in northeast Thailand, planners (by the set up of the project itself) favoured conventional knowledge provided by state actors, and as such violated one core principle associated with communicative planning: that for collaborative planning there are no privileged types of knowledge (Brand and Gaffkin, 2007).[10] This, however, would have greatly improved both the quality of the generated plans and the deliberation at the workshop.

Also, collaboration is often presented as the only valid stakeholder strategy to resolve differences. However, and particularly in the context of northeast Thailand, this neglects the fact that stakeholders and civil society organizations might have important other strategies available to open up political space and democratize planning. As Thai political scholar Somchai Phatharathananunth (2006) pointed out, 'the importance of civil society for democratic development in Thailand is more pronounced because of the exclusionary nature of Thai democracy'. In addition, the author argued that 'only through political mobilization can movements build their bargaining powers and force the state to recognize them as political forces unable to be ignored'.

Of course, the above should not suggest that collaborative planning (at least in theory) does ignore the role of power in planning (see for example Healey, 2003). Dore (2007), for example, argued that power relationships embedded in the Mekong region political context undoubtedly influence the extent that meaningful participation and negotiation is possible. However, there is evidence that powerful actors have increasingly co-opted the initial hypothesis and rhetoric, which now more often than not masks business-as-usual approaches to planning. This confirms Edmunds and Wollenberg (2002) who argued that powerful groups often manipulate seemingly neutral terms that are quickly agreed to in meetings, but then use them in ways that meet their very own needs. At the same time, the politics at work are scarcely analysed or discussed, as water resources planning documents are thoroughly de-politicized and invariably given a techno-centric veneer.

Conclusion

Official documents on IWRM principles and practices, both within Thailand and the Mekong region more generally, often suggest that there have been tangible shifts in planning and policy paradigms. However, there is a wide gap between the rhetoric adopted both in national and international mainstream publications advocating better planning practices and the real-politics of water resources planning in the Lower Mekong Basin (see also Molle, 2007; Sneddon and Fox, 2007).

We have argued that at the same time that collaborative modes of planning are becoming mainstreamed, the initial assumptions are increasingly obliterated or universalized to mask complexity, as actual implementation more closely resembles business-as-usual practices. This suggests that there is a likelihood that the catchphrases of participatory planning, collaborative planning, dialogue and negotiation, which today take centre stage in international policy and planning recommendations, will witness a similar fate as previous normative buzzwords and concepts that have preoccupied development scholars and practitioners, such as 'sustainable development', 'trade-offs' (see Friend and Blake, 2009) and also 'IWRM' itself.

Flyvbjerg (2002) has argued that communicative planning fails to capture the role of power in planning. As evidenced in the case of the planning for the water transfer project under the Mekong Water Resources Assistance Strategy, deliberation was initially confined to 'internal' stakeholders, which frustrated the participation of the wider (and active) Thai civil society. In turn, those excluded had to use alternative strategies to open up a limited political space in order to voice their concerns within the planning efforts. The struggle for simple rights to be heard in nominally the most democratic of the Lower Mekong Basin nations, let alone rights to water resources and participation in water control negotiations, come starkly into focus in this instance.

Of course, the struggle and slow pace to move from passive participation to more substantial stakeholder involvement and collaborative planning is not confined to northeast Thailand or the Mekong basin. Wester et al (2003) have made similar observations in Mexico and South Africa. Yet, Flyvbjerg's (1998, 2002) argument that a renewed focus on 'what is actually done', offers better prospects than focusing on 'what should be done'. In the words of the author: 'If the goal of planning theorists is to create societal change which is closer to Habermas' ideal society – free from domination, more democratic, a stronger civil society – then the first task…is to understand the realities of power' (Flyvbjerg, 2002). This is supported by Edmunds and Wollenberg (2002) who argued that negotiations, deliberation and participation will achieve more 'if we are more open in discussing the politics at work', and also confirms Leal (2007) who argues that 'participation needs to be re-articulated within broader processes of social and political struggle'. Ultimately, the inherently political nature of water resources planning, management and control underpins all decisions and is hard to escape in the real world.

The case presented the seemingly unbridgeable gap that appears between official rhetoric and practice in the field of water resources development, once the messy reality of everyday politics is put under the spotlight in the highly contested waterscapes of the Mekong Basin. Despite the over-exuberant praise heaped on Thailand as a regional leader in participatory approaches to water resources management in its river basins, this case questions the actual implementation of this concept. Perhaps it is still early days in seeing how the experiment in IWRM principles will actually play out and be delivered on the ground even as Thailand continues to undergo a period of political instability and reform, which may also present new opportunities for previously poorly represented actors to genuinely participate in water resources decision-making processes at the river basin and national levels.

Acknowledgements

Research leading to this paper was supported by the Challenge Program on Water and Food, with funding support from the European Community via the International Fund of Agricultural Development. The authors want to express their gratitude to François Molle, John Dore and Richard Friend for constructive comments on an earlier draft, as well as numerous M-POWER colleagues for the discussions leading to it.

NOTES

1 River Basin Management 'is a more traditional term which has recently broadened its meaning to encompass many of the same features and values which characterize IWRM' (Svendsen et al, 2005).
2 For a more complete history of irrigation planning and development in northeast Thailand see for example: Chomchai, 1994; Sneddon, 2003; Floch et al, 2007; Molle and Floch, 2008; Molle et al, 2009.
3 In its original incarnation, the dam designed by the US Army Corps of Engineers would have an installed generation capacity of 4800MW and required the relocation of 280,000 people in its reservoir footprint, but was later scaled down in size, before being eventually abandoned in the 1990s as unrealistic.
4 The three sub-regional areas identified included: 1) the sub-region shared by Thailand and Lao PDR along the Mekong River; 2) the 3S area (Sekong, Sesan and Srepok) shared by Cambodia, Lao PDR and Vietnam; and 3) the parts of the Mekong Delta shared by Vietnam and Cambodia (WB and ADB, 2006).
5 Importantly, this included options for irrigated agricultural development in the Vientiane Plain and the Khammouane-Xe Bangfai Plain in Lao PDR.
6 Apart from the initial estimates with regard to benefits to Laos, the Nam Ngum River Basin Integrated Water Resources Management Plan stated that 'the proposal include the opportunity costs if the water could be used for irrigation or other purposes in Laos, the possible extensive flooding of the Vientiane Plain from the barrage, a barrier to fish passage from the Mekong to the Nam Ngum Basin rivers, water security to existing and future users, water royalties and impacts for existing and possibly new tourism development' (WREA, 2009). Also, the report (p52) classified the risk of negative impacts from out of basin transfer as 'High', with potential 'highly negative impacts' with regards to the environment, fisheries and flooding. Importantly, the report does not indicate one single category that would result in positive impacts.
7 The River Basin Committee (RBC) concept is a product of the rhetorical adoption of IWRM principles at the national level that emerged from the formulation of a national water vision and policies following the creation of the Department of Water Resources in 2002.
8 Composed of Japanese private sector consultants, accompanied by Khon Kaen University academics engaged to prepare the Thai components under the study.
9 'Khong' refers to all the tributaries of the Mekong River lying in northeast Thailand, excluding the Chi and Mun river basins, which arbitrarily forms one of 25 identified river basins lying within Thai territory and used to demarcate an imagined hydro-ecological unit.
10 Within northeast Thailand, alternative forms of knowledge production have evolved over the last 20 years, including Thai Baan research.

REFERENCES

ADB (2003) 'Water for all: The water policy of the Asian Development Bank', The Asian Development Bank, June 2003, Manila, The Phillippines

ADB (2004) 'Thailand: A regional leader in water resources management', The Asian Development Bank, August 2004, Manila, The Phillippines
AMRC (2007) 'Unpacking the Mekong water resources assistance strategy', Mekong Brief Number 6, September 2007, Australian Mekong Resources Centre, University of Sydney, Sydney, Australia
BIC (2005) 'The Greater Mekong Subregion and MDBs: Basic facts and updates', GMS Update #1, Bank Information Center, available at www.bicusa.org/Legady/GMS_Update1.doc (last accessed October 2009)
Blake, D.J.H. and Floch, P. (2009) 'Conflated problems and solutions? Water resources crisis narratives in Northeast Thailand and the infrastructure development imperative', Association of South-East Asian Studies in the United Kingdom (ASEAUK), 11–13 September 2009, Swansea University, UK (unpublished draft)
Brand, R. and Gaffkin, F. (2007) 'Collaborative planning in an uncollaborative world', *Planning Theory*, vol 6, no 3, pp282–313
Chang Noi (2009) *Jungle Book: Thailand's Politics, Moral Panic and Plunder 1996–2008*, Silkworm Books, Chiang Mai, Thailand
Chomchai, P. (1994) 'The United States, the Mekong Committee and Thailand: A Study of American multilateral and bilateral assistance to Northeast Thailand since the 1950s', Asian Studies Monograph No 051, Institute of Asian Studies, Chulalongkorn University, Bangkok
Consultant Team (2006) 'Scoping the options for joint water resources development and management between Lao PDR and Thailand in the Mekong Basin: Inception report', June, Shuhei Seyama for the World Bank
Consultant Team (2007a) 'Scoping the options for joint water resources management between Lao PDR and Thailand in the Mekong River Basin: Second working paper – Outline of options and issues', consultant report prepared for the World Bank and the Mekong River Commission Secretariat, February 2007
Consultant Team (2007b) 'Participants and comments: Thai provincial workshop, 23 February 2007', scoping the options for joint water resources management between Lao PDR and Thailand in the Mekong River Basin. Khon Kaen, Thailand
Cooke, B. and Kothari, U. (eds) (2001) *Participation: the New Tyranny?* Zed Books, London and New York
Dore, J. (2007) 'Multi-stakeholder platforms (MSPs): Unfulfilled potential', in L. Lebel, J. Dore, R. Daniel and Y.S. Koma (eds) *Democratizing Water Governance in the Mekong*, Mekong Press, Chiang Mai, Thailand, pp197–226
Edmunds, D. and Wollenberg, E. (2002) 'Disadvantaged groups in multistakeholder negotiations', CIFOR Programme Report, June 2002, Centre for International Forestry Research, Bogor Barat, Indonesia
Ekachai S. (2008) 'Descructive mega dreams', *Bangkok Post*, 12 June 2008
ESCAP (2005) 'Good practices on strategic planning and management of water resources in Asia and the Pacific', Water Resources Series No 85, Economic and Social Commission for Asia and the Pacific, United Nations, Bangkok, Thailand
Fainstein, S.S. (2000) 'New directions in planning theory', *Urban Affairs Review*, vol 35, pp451–478
Floch, P., Molle, F. and Loiskandl, W. (2007) 'Marshalling water resources: A chronology of irrigation development in the Chi-Mun Basin, Northeast Thailand', M-POWER Working Paper, Chiang Mai

Flyvbjerg, B. (1998) 'Habermas and Foucault: Thinkers for civil society?', *British Journal of Sociology*, vol 49, no 2, pp210–233

Flyvbjerg, B. (2002) 'Bringing Power to planning research: One researcher's praxis story', *Journal of Planning Education and Research*, vol 21, pp353–366

Forester, J. (1980) 'Critical theory and planning practice', *Journal of American Planning Association*, vol 46, no 3, pp275–286

Friend, R.M. and Blake, D.J.H. (2009) 'Negotiating trade-offs in water resources development in the Mekong Basin – Implications for fisheries and fishery-based livelihoods', *Water Policy*, vol 11, S1, pp13–30

Gujit, I. and Shah, M.K. (1998) 'Waking up to power, conflict and process', in I. Gujit and M. Shah (eds) *The Myth of Community: Gender Issues in Participatory Development*, Internmediate Technology Publications, London, pp1–23

GWP (2000) 'Integrated water resources management', TAC (Technical Advisory Committee) Background Paper No 4, Global Water Partnership, Stockholm

Healey, P. (2003) 'Collaborative planning in perspective', *Planning Theory*, vol 2, no 2, pp101–123

Heyd, H. and Neef, A. (2004) 'Participation of local people in water management: Evidence from the Mae Sa Watershed, Northern Thailand', EPTD Discussion Paper No 128, International Food Policy Research Institute, Washington DC

Huxley, M. (2000) 'The limits to communicative planning', *Journal of Planning Education and Research*, vol 19, pp369–377

ITD (2007) 'The 10th National Economic and Social Development Plan's feedbacks by Saralthorn Tanomsup', International Institute for Trade and Development, available at www.itd.or.th/en/node/308 (last accessed October 2009)

Lachapelle, P.R., McCool, S.F. and Patterson, M.E. (2003) 'Barriers to effective natural resources planning in a "messy" world', *Society and Natural Resources*, vol 16, pp473–490

Leal, P. A. (2007) 'Participation: The ascendancy of a buzzword in the neo-liberal era', *Development in Practice*, vol 17, no 4, pp539–548

Lim, G.-C. (1986) 'Toward a synthesis of contemporary planning theories', *Journal of Planning Education and Research*, vol 5, pp75–85

Middleton, C. (2007) 'The ADB/WB/MRC "Mekong Water Resources Assistance Strategy": Justifying large water infrastructure with transboundary impacts', paper prepared for Critical Transitions in the Mekong Region, 29–31 January 2007, Chiang Mai, Thailand

Molle, F. (2007) 'Irrigation and water policies: Trends and challenges', in L. Lebel, J. Dore, R. Daniel and Y.S. Koma (eds) *Democratizing Water Governance in the Mekong Region*, Mekong Press, Chiang Mai, Thailand

Molle, F. (2008) 'Nirvana concepts, narratives and policy models: Insights from the water sector', *Water Alternatives*, vol 1, no 1, pp131–156

Molle, F. and Floch, P. (2008) 'Megaprojects and social and environmental changes: The case of the Thai "Water Grid"', *AMBIO*, vol 37, no 3, pp199–204

Molle, F., Floch, P., Promphaking, B. and 2Blake, D.J.H. (2009) '"Greening Isaan": Politics, ideology, and irrigation development in Northeast Thailand', in F. Molle, T. Foran and M. Käkönen (eds) *Contested Waterscapes in the Mekong Region: Hydropower, Livelihoods and Governance*, Earthscan, London

Mollinga, P.P. (2008) 'Water, politics and development: Framing a political sociogy of water resources management', *Water Alternatives*, vol 1, no 1, pp7–23

Mouffe, C. (1999) 'Deliberative democarcy or agonistic pluralism', *Social Research*, vol 66, no 3, pp745–758

MRC (2005) 'Public participation in the Lower Mekong Subregion', Mekong River Commission, Vientiane, Lao PDR

Neef, A. (2009) 'Transforming rural water governance: Towards deliberative and polycentric models?', *Water Alternatives*, vol 21, no 1, pp53–60

NESDB (1996) 'The Eighth National Economic and Social Development Plan (1997–2001)', National Economic and Social Development Board, Bangkok, Thailand

NESDB, UNEP and TEI (2008) 'National Sustainable Development Strategy for Thailand: A Guidance Manual. National Economic and Social Development Board', United Nations Environment Programme and Thailand Environment Institute, Bangkok, Thailand

Novotny, H. (1999) 'The place of people in our knowledge', *European Review*, vol 7, no 2, pp247–262

Özerol, G. and Newig, J. (2008) 'Evaluating the success of public participation in water resources management: Five key constituents', *Water Policy*, vol 10, pp639–655

Phatharathananunth, S. (2006) *Civil Society and Democratization: Social Movements in Northeast Thailand*, NIAS Press, Copenhagen

Pheddara, P. (2007) 'Large rice-based irrigation systems in Lao PDR', in Z. Chen (ed.) *Future of Large Rice-based Irrigation Systems in Southeast Asia*, proceedings on the Regional Workshop on the Future of Large Rice-based Irrigation Systems in Southeast Asia, Ho-Chi Minh City, Vietnam, 26–28 October, 2005, FAO Regional Office for Asia and the Pacific, Bangkok, pp101–107

Protest Communiqué (2007) 'Anti-water transfer between Thailand and Laos: Announcement'(original in Thai), 27 February, Khon Kaen, unpublished

Rogers, P. and Hall, A. (2003) 'Effective water governance', GWP Technical Committee Background Paper 7,Global Water Partnership, Stockholm

Saarikoski, H. (2002) 'Naturalized epistemology and dilemmas of planning practice', *Journal of Planning Education and Research*, vol 22, pp3–14

Samabuddhi, K. (2004) '"Water crisis looms", says grid study', *Bangkok Post*, 13 June

Sanyu Consultants Inc. (2004) 'Nam Ngum Water Management Project for Vientiane Plain of Lao PDR and Northeast Thai Region', Sanyu Consultants Inc. in association with Sanyu Consultants (Thailand) Ltd, February

Sethaputra, S., Thanopanuwat, S., Kumpa, L. and Pattanee, S. (2001) 'Thailand's water vision: A case study', in L.H. Ti and T. Facon (eds) 'From Vision to Action: A Synthesis of Experiences in Southeast Asia', The FAO-ESCAP Pilot Project on National Water Visions, Environment and Natural Resources Development Division of ESCAP and the FAO Regional Office for Asia and the Pacific, Bangkok, Thailand, pp71–97

Sneddon, C. (2003) 'Reconfiguring scale and power: The Khong-Chi-Mun project in northeast Thailand', *Environment and Planning*, vol 35, pp2229–2250

Sneddon, C. and Fox, C. (2007) 'Power, development and institutional change: Participatory governance in the Lower Mekong Basin', *World Development*, vol 35, no 1, pp2161–2181

Southeast Asia Rivers Network (2002) 'Laos-Thai Friendship Water Diversion Project', excerpt from the Conceptual Study Report on Laos-Thai Friendship Water Development for Sustainable Agriculture in Savannakhet Province of Lao PDR and Lower Chi Basin of Thailand, June 1998, conducted by Sanyu Consultants Inc., prepared by Southeast Asia Rivers Network, Thailand Chapter, 20 June

Svendsen M., Wester, P. and Molle, F. (2005) 'Managing river basins: An institutional perspective', in M. Svendsen (ed.) *Irrigation and River Basin Management: Options for Governance and Institutions*, CABI Publishing and International Water Management Institute, Wallingford, UK, and Sri Lanka

Thanopanuwat, S. (2008) 'Northeast Thailand water shortage and approach to sustainable use of Mekong Water', in proceedings of the 3rd International Symposium on Sustainable Development in the Mekong River Basin, Khon Kaen, Thailand. 11–12 September, 2008, Japan Science and Technology Association and Royal Irrigation Department, Bangkok, Thailand, pp101–108

The Hague Ministerial Declaration (2000) 'Ministerial Declaration of the The Hague on Water Security in the 21st Century', 22 March 2000, The Hague, The Netherlands, available at www.gdrc.org/uem/water/hague-declaration.html (last accessed 14 November 2010)

Ti, L.H. and Facon, T. (2001) 'From vision to action: A synthesis of experiences in Southeast Asia', The FAO-ESCAP Pilot Project on National Water Visions, Environment and Natural Resources Development Division of ESCAP and the FAO Regional Office for Asia and the Pacific, Bangkok, Thailand

UNDP (2006) 'Multi-Stakeholder Engagement Processes: A UNDP Capacity Development Resource', Conference Paper #7, Working Draft, Capacity Development Group, Bureau of Development Policy, United Nations Development Programme

USAID (2007) 'What is integrated water resources management?', United States Agency for International Development, available at www.usaid.gov/our_work/environment/water/what_is_iwrm.html (last accessed July 2009)

Wester, P., Merry, D.J. and de Lange, M. (2003) 'Boundaries of consent: Stakeholder representation in river basin management in Mexico and South Africa', *World Development*, vol 31, no 5, pp797–812

Wipatayotin, A. (2008) 'Concern over Laotian water diversion plan: Samak's pet project requires EIA study', *Bangkok Post*, 19 July

WB and ADB (2006) WB/ADB Joint Working Paper on Future Directions for Water Resources Management in the Mekong River Basin: Mekong Water Resources Assistance Strategy (MWRAS), June, available at www.adb.org/water/operations/partnerships/mwras-June2006.pdf (last accessed October 2006)

WREA (2009) 'Nam Ngum River Basin Integrated Water Resources Management Plan', Water Resources and Environment Administration, March 2009, Vientiane, Lao PDR

3

Local People's Participation in Involuntary Resettlement in Vietnam: A Case Study of the Son La Hydropower Project

Tran Van Ha

INTRODUCTION

The economic reforms under *Doi Moi* (decreed in 1986 at the Sixth National Congress) have brought new demands for energy to drive Vietnam's economic development. The Government of Vietnam has identified construction of hydropower dams as a key strategy to meeting these energy demands. During the last 15 years, the country has experienced a surge in hydropower construction led by Electricity of Vietnam (EVN), the largest power company, followed by other state and national companies. The social impacts of large dam construction have become increasingly well-known in the global debate on sustainable development (Cook, 1993; Cernea and McDowell, 2000). In recent years, with the establishment of the Vietnam Rivers Network (VRN), a number of studies on resettlement in Vietnamese hydropower projects have been carried out by VRN's members (Eco-Eco, 2004; Doan and Nguyen, 2006; Hoang, 2006; Vo, 2006; Dao, 2006; Tran and Le, 2008). Since most dams are located in mountainous areas, these studies focus on resettlement in different mountainous regions: in the northwest (VUSTA, 2006; Hoang and Vo, 2006; Tran and Le, 2008), in the central region (Doan and Nguyen, 2006; Hoang, 2006; Vo, 2006) and in the central highlands (Dao, 2006). Other studies focus on the effects of resettlement and sustainable development in mountainous villages (Pham, 1995).

Demonstrating the high level of interest among both Vietnamese and international scholars, these studies have provided valuable insights into resettlement and its impacts on stabilizing ethnic communities' lives and livelihoods in areas of low socio-economic development. Local communities, directly or indirectly affected by the construction of dams in the mountainous regions, are often ethnic minorities, which means that resettlement programmes must take into consideration the rich diversity of traditional practices, social organizations and knowledge systems.

However, most of these studies have not systematically analysed the participation of local people in resettlement programmes. This chapter aims to explore the processes through which local people participate in resettlement programmes. This examination is carried out in the context of how hydropower projects impact the roles, rights and decision-making processes of people in local communities in ways that affect their livelihoods. As will be discussed below, besides constraints in the government policies implementation, there are many barriers to effective participation of upland peoples, including their cultural differences, geographic and topographic conditions, capacity of local authorities and varying levels of demand for participation at different levels of implementation.

This research is based on two field surveys in the northern mountains of Vietnam that are affected by the construction of the Son La Dam, which is expected to be the largest dam in Southeast Asia when it is completed by the end of 2010. The first field survey, conducted in 2005–2006 in Son La and Lai Chau provinces, involved 25 affected communities, including both resettled and host communities. The second, carried out in 2008, covered eight resettled villages in the same provinces. The research methods included questionnaire-based, open-ended interviews with villagers, key informants at village, commune and district levels, and resettlement officers, combined with government materials and policy analysis. This chapter aims to examine changing roles and responsibilities of dam-associated resettled communities.

The structure of the chapter is as follows: the first section discusses general concepts associated with participation and policy developments in the enabling legal and regulatory environment in Vietnam. The second section explores the processes of local participation in the resettlement programme of the Son La Dam. In the third section, there is a discussion of the implications of the Son La experience for implementation of participation in other resettlement projects in Vietnam. The concluding section reflects on the challenges of improving people's participation in dam resettlement projects in the future.

Participation in Vietnam: Development of Key Concepts

Analysis of participation often boils down to an examination of the gaps between development policy and practice, or perhaps a more critical appraisal of the reality of participation within specific project contexts. Large-scale infrastructure development, and particularly water resources development, has been an especially productive arena for looking at these gaps. In Vietnam, the period between the construction of the Hoa Binh Dam in the 1970s and the Son La Dam in the 2000s saw significant improvements in the supporting institutional framework under which participation is carried out (Dao, 2010). The first experiences with resettlement occurred within a weak legal framework, without the benefit of international best practices to provide a benchmark to assess performance. Currently, the integration of Vietnam's development into global governance processes, such as the World Commission on Dams, has meant that not only has the policy and legal framework been improved, but lessons learned from global best practice can be applied. Public debate, led by NGOs and the media, has brought more of the resettlement processes, including the participation aspects, under closer scrutiny by Vietnamese society.

Principles of participation are often set in the global debate on environment, development and human rights, but international organizations, development agencies and donor countries have differing approaches to promoting participation. At the same time, participation is implemented within the legal, regulatory and socio-economic context of each country and region. Nevertheless, there is a certain urgency in the search for governance mechanisms to close these gaps, as they have both direct and indirect impacts on local communities – direct in terms of affecting their socio-economic and cultural well-being, indirect as the scope of their long-term rights and responsibilities in development decision-making is determined. While this section does not provide a comprehensive analysis of the vast body of participation literature, it focuses on how participation has fared in the implementation of the Son La resettlement programme.

Participation and resettlement in Vietnam

Participation is a relatively new concept in Vietnam. However, in the 1950s, as part of the post-colonial nation-building process, the Prime Minister's Decree No. 151/1959/NĐ-TTg on regulations on recovering land referred to the involvement of 'grassroots' people in state intervention programmes. Since then, the concept has come to provide a role for local people in economic development and other aspects of governance in Vietnam. The evolution of Vietnamese ideas of participation has also been influenced by the global discourse in more recent years (Diep, 1995). For example, a key component of participation is empowering people to be 'masters of their fate', which is based on the belief that democratic society at large may benefit from people's participation in development processes. At the same time, development cannot be sustainable unless people have a central role in critical aspects of decision-making (Kumar, 2002, cited in Le and Nguyen, 2006). In the water sector, resettlement programmes provide an important arena of decision-making in which local people's participation is necessary. According to the Asian Development Bank (ADB, 1992), an important source of hydropower development finance for Vietnam, consulting affected people is the initial point of entry for every activity that involves resettlement (Pham, 1997). Effective participation in resettlement processes addresses the affected people's concerns and offers them opportunities to take part in the decisions that will affect their lives (Diep, 1995). As Duong (2009) argues, full information disclosure at the onset of a project will limit misunderstanding and increase trust, which lays the foundation for cooperation between the affected people and project proponents. Moreover, resettlement without people's participation may lead to unsuitable rehabilitation strategies and impoverishment of the affected people and increase problems later if not dealt with at the onset. With regards to resettlement, experience has shown that participation of resettled communities and host communities are both important if negative impacts are to be reduced (Diep, 1995; Pham, 1995; Duong, 2009).

The revised law on environmental protection in 2005 is an improvement in government policy in facilitating participation of local people in resettlement projects associated with hydropower construction. The law requires a Strategic Environmental Assessment (SEA) when the investor designs a hydropower project. According to the law, an SEA includes not only an assessment of environmental impacts but also consideration of negative socio-economic and cultural impacts that may occur as a result of the project's implementation. Therefore, the law allows local people to participate in the design and planning processes, enabling them to raise their voices on essential issues such as water supply, livelihood security and socio-economic sustainability, as well as cultural preservation. The law also stipulates that a hydropower project may not be implemented unless its SEA (including the resettlement plan) has been approved by the relevant authorities. However, there is still a lack of clear mechanisms for people to participate. There is also a need for an independent monitoring system to ensure that non-biased, accurate and sufficient information will be collected in the SEAs.

In general, Vietnam still lacks technical capacity and financial resources to implement SEAs for all of its projects. The implementation of SEA thus far has not been effective in terms of facilitating meaningful participation in dam-associated resettlement projects. The Son La Dam was initiated before approval of the amended law on environmental protection, so the SEA requirement was not applicable to the resettlement process itself. Rather, an assessment of the environmental impacts of the resettlement project was implemented. The final approval of the environmental impact assessment (EIA) was given in 2007, almost two years after the construction of the dam had started. The remaining question is whether the lessons are incorporated into the decision-making process of future proposed dams to enable meaningful participation in this strategic decision-making step.

Ethnic minorities and participation

It is widely recognized that ethnic minority communities are often marginalized and otherwise adversely affected by large-scale development projects (VUSTA,

2006; Dang, 2007; Dao, 2010, Cariño and Colchester, 2010). One central reason for this worrying trend is the lack of effective mechanisms for participation in the planning and implementation at key stages in the project (Cook, 1993; Ribot, 1998). In response to this, a global indigenous peoples' movement has been working to increase indigenous peoples' rights to free, prior and informed consent (FPIC), which is clearly recognized by organizations such as the United Nations (UN). The Declaration on the Rights of Indigenous Peoples (UNDRIP) was adopted by the UN General Assembly in 2007 and articulates indigenous peoples' rights to FPIC[1] (UNDRIP, 2007 in Cariño and Colchester, 2010). The Declaration sets out what it describes as the minimum standards for the survival, dignity and well-being of the indigenous peoples of the world and codifies a series of existing norms regarding indigenous peoples (Cariño and Colchester, 2010). However, how this is articulated in practice in a country like Vietnam where movements of peoples, particularly ethnic minorities, are rather weak is an interesting area to be explored.[2]

In Vietnam, there are differences in participation between hydropower-induced resettlement projects in mountainous areas, where ethnic minorities make up a large portion of local communities, and other development projects in lowland and urban areas that are demographically dominated by the Kinh (ethnic Vietnamese). The history, customs and cultures of ethnic minorities in mountainous areas, which influence the communities' and individuals' roles as well as participation rights, differ from those of the Kinh people in lowland areas. Research by anthropologists and sociologists has shown that upland people have long exercised their rights to participate through communities' customary laws (Cam and Ngo, 1999). These rights, however, were founded on specific cultural practices of local ethnic groups, and did not go beyond the border of their communities or kinship relationship. Importantly, they were not recognized by the formal legal system either. The Thai peoples' Sen ban (annual worship ceremony for health and peace) is a good example of local participation. Before the ceremony the entire village selected one person to be in charge and lead the annual worship ceremony. This person was chosen from one of the first families who created the village. In addition, this selected person had to discuss how he wanted to organize the ceremony with the Tao ban (headman) and the Thau Ke (elderly council) and obtain consensus at the meeting. After the meeting, Chau viec (the messenger) of the village was responsible for delivering information to every single family. This information included the time and procedure for the ceremony, how it was to be organized and the responsibility of each family. The whole village worked together in preparing for the activity. People believed that since every family participated before and during the ceremony, they all would receive support from superior forces.

In most hydropower projects in Vietnam implemented between the 1960s and early 1990s, resettlement was not regarded as an important component of planning. There were no clear regulations on the procedures for setting up and assessing resettlement plans from the preparation stage (Dao, 2010). There was a lack of coordination among the concerned agencies, and a distinct absence of participation and consultation with both resettled and host communities (Pham and Lam, 2000). For example, while resettling local people from the Thac Ba reservoir in the 1960s and the Hoa Binh reservoir in the 1970s and 1980s, rehabilitation of ethnic minority communities was not among the immediate concerns of the agencies promoting the projects (Diep, 1995).

In brief, resettlement projects in Vietnam have not paid enough attention to social and cultural differences of ethnic minorities, which contributes to restraining people's ability to participate in projects that have caused adverse impacts on their lives. Therefore, it is important to examine the evolution of the Vietnamese government's resettlement policy over the last few decades.

Policy and practice: Identifying the gaps

By considering how policies have evolved we can help to distinguish the limitations of the resettlement concept in development projects in general and hydropower construction in particular, both in terms of communities' and individuals' participation rights and roles. Resettlement is relatively common in Vietnam now but there is great variation in the levels of concern about local peoples' participation in stabilizing their own lives and livelihoods. Before 1993, the government approved development projects without explicit resettlement plans, which were undervalued and regarded as minor part of projects. Little concern was shown for the living conditions of those who were actually resettled by the new dam, nor were they given the opportunity to participate in planning for their own resettlement.[3]

Since 1993, the government has acknowledged the importance of resettlement plans and the need to ensure the enactment of relevant policies. Lessons from the Hoa Binh hydropower project have shown the government that if they did not implement resettlement programmes properly, they would have to address many subsequent problems both in the short and long term. Thus, the government requested that relevant state institutions broaden their viewpoints so that not only construction aspects of projects are considered but also people's livelihoods. The authorities were obliged to take more responsibility to reduce local peoples' losses through measures to improve the affected people's lives. As a result, a number of decisions and decrees related to resettlement have been issued, such as Decree 22/1998/NĐ-CP, which identifies the subjects to be compensated and Decree No 197/2004/NĐ-CP on 'Compensation, Assistance and Resettlement When the State Recovers Land', among others.

Globally there has been much discussion about norms of both the policy and practice of participation. As influential players on the international development stage, the World Bank (WB) and Asian Development Bank (ADB) have made a contribution to advancing the agenda of local people's participation in development activities, through the Banks' own internal policies and support to national policy frameworks. However, both Banks have come under serious scrutiny, particularly over the contradictions between their own gaps in policy and practice especially with regards to investments in large dams (see Hirsch, 2001; Middleton et al, 2009; Singh, 2009). Regardless of the debate on the effectiveness of the Banks' participation, they have been influential in providing a facilitating environment for deeper examination of participation in Vietnam.

Although greatly improved since 2000, there is still a remarkable gap between the right to participate by affected people in resettlement policies of Vietnam and those proposed by international organizations such as the WB and ADB.[4] For example, according to WB and ADB regulations, affected people should be given full information on policies with regards to compensation, support and restoration and consulted during planning and implementation processes. In contrast, according to the Land Law of Vietnam, 1993,[5] Article 28 includes that the Government of Vietnam only needs to announce a project and not consult with stakeholders. 'Prior to expropriating the land, the land users need to be informed of the reasons, timing, displacement plan, and compensation solution.' However, Decree 22, which was approved in 1998, requires a representative of affected peoples in the resettlement compensation board, which represents a significant improvement in people's participation in resettlement projects directly related to them (Pham and Lam, 2000).

A gap between Vietnamese and international policies is also seen in the supervision of resettlement projects. In accordance with WB and ADB regulations, in addition to internal supervision, project implementers must employ external organizations to independently supervise resettlement to assess the effectiveness of the target achievement and the socio-economic effects on resettled people. On the contrary, independent supervision is not required in Vietnamese resettlement projects (Pham and Lam, 2000). However, at a higher level of consideration, these gaps may be more a product of the types of norms that are being advanced. 'Global norms' promoted by international development agencies are similar to government norms in that they are disconnected from the traditional practices and socio-economic conditions of local communities. As discussed briefly above, local groups often have traditions and institutions that facilitate appropriate participation in local matters. Incorporating local norms of participation, confidence building, communication and decision-making into development projects would probably lead to rather different results. Indeed, there is substantial literature from other places supporting the conclusion that participation led by state and international agencies is often blind to existing forms of local participation, and is discussed further in the following sections (Cooke and Kothari, 2001).

This brief review of Vietnam's experiences with participation in resettlement programmes and hydropower development has uncovered two sets of gaps between policy and practice. The first gap is between accepted international practice, as seen in both the multilateral development banks and the Vietnamese legal and policy framework. The second gap is between the stated Vietnamese procedures, as stimulated in laws and regulations, and the practice on the ground.

THE SON LA HYDROPOWER PROJECT: RESETTLEMENT AND CHALLENGES TO PARTICIPATION

The Son La Dam project provides a useful window on the current directions in participation policy and practice in Vietnam. Because of its scale, the government's handling of the project is setting the policy precedent for resettlement programmes in the country, which will have long-term implications for the basic framework for water governance in the future. This section addresses two main challenges that emerged in the process of implementation. First is that resettlement plans as originally articulated were not meeting the expectations of local people. Furthermore, conflicts among local communities were threatening the feasibility of the resettlement programmes. Second, it soon became clear that the outcomes of implementation varied between the Son La and Lai Chau provinces involved in the project, meaning that the foundation for public participation in the programme was not the same. This affected the level of participation that was achieved by the affected people.

The Son La Dam: Rationale and challenges

The Son La Dam, upon its completion at the end of 2010, will be the largest dam in Southeast Asia, with an installed capacity of 2400MW and the ability to generate around 14 billion kWh annually (Son La Province's People's Committee, 2006). The Son La Dam is expected to contribute to economic growth in Vietnam by increasing the supply of low cost energy to meet rising demand from urban and industrial growth and rural development. It is estimated that 75 per cent of the electricity production from the plant will serve the northern region and the remaining 25 per cent will serve the central and south regions of the country. The dam proponents argue that without this project, there will be a shortage of power in northern Vietnam due to the limitation of gas and coal resources. The Son La dam is a multi-purpose project. So, besides power generation, the project will provide other benefits such as flood control and irrigation.

However, besides the above-mentioned potential benefits, the construction of the dam also has negative impacts. The total area of land submerged by the reservoir is approximately 23,000 hectares, of which 7600 hectares is agricultural and 3200 hectares is forest. Specifically, land that will be lost due to the Son La project includes riverbank gardens, fertile fields for rice and subsidiary crops and forests (NIAPP, 2004).

The reservoir will also submerge 248 villages in 31 communes in eight districts and towns of Son La, Lai Chau and Điện Bien provinces (see Figure 3.1). Thus a great concern related to the building of this dam is the resettlement of approximately 18,897 households in these provinces.[6] These resettled people belong to 10 different ethnic groups, of which Thai people (Thai Den and Thai

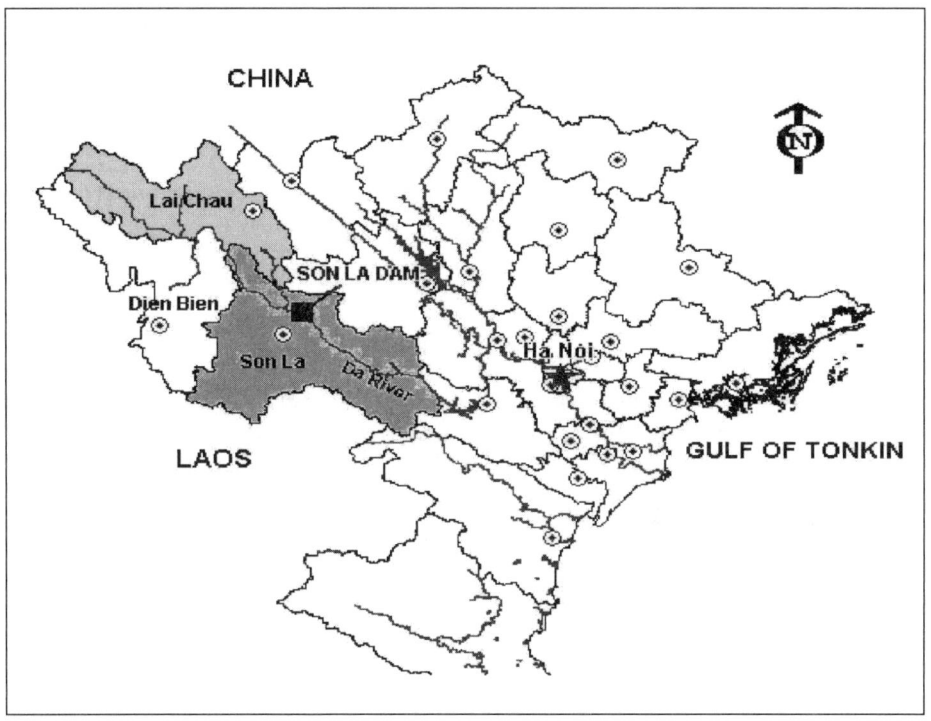

Figure 3.1 *Map of Son La Dam location and field study sites*

Source: VUSTA, 2006

Trang) comprise 74 per cent, Kinh (national majority) account for 11 per cent and the remaining 15 per cent are ZiaoMien, La Ha, Kho Mu and Khang (Institute of Geography, 2005). This means that 89 per cent of the affected people are ethnic minorities, belonging to four of the five major ethno-linguistic groups of mainland Southeast Asia – Tai-Kadai, Sino-Tibetan, Mon-Khmer and Hmong-ZiaoMien. An effective resettlement programme for this dam thus needs to accommodate this tremendous range of human diversity expressed in differing agricultural systems, ritual practices, forms of social organization and customary knowledge and values, not to mention the necessity for providing information and eliciting inputs from people speaking languages that are not understood by the government agencies.

Adjusting the resettlement policy to be more responsive

The scale of the Son La Dam and its resettlement needs have been met with specific policy adjustments. The government issued Decree No. 197/2004/NĐ-CP (Government of Vietnam, 2004b) on guidance for compensation, assistance and resettlement and also the prime minister's Decision No. 459/QĐ-TTg on compensation and resettlement (Government of Vietnam, 2004a) and Decision No. 196/2004/QĐ-TTg on details of budgetary resources for resettlement (Government of Vietnam, 2004c).

However, after a couple of years of applying Decision No. 459 to the resettled communities under the 145 metre level,[7] two major problems emerged. First, since resettled people had to move into existing communities in designated resettlement areas, policy adjustment to assist host communities with running water and electricity were needed. In some places, there were difficulties in appropriating land from the host communities for resettlement sites. Host communities, whose lands were acquired and who had not yet received any compensation or benefits from basic infrastructure such as running water and electricity, felt that the project was unfair to them. In fact, in some areas, where host communities used to receive support from the government for being in especially difficult and remote mountainous areas,[8] the authorities decided that these people were no longer eligible to receive special support since they had already received benefits from the resettlement programme. This resulted in great dissatisfaction by the majority of the communities. Second, in the resettlement sites, resettled people had not received adequate support for agricultural production when they moved to the new areas. Even in 2008, people who were moved to resettlement sites in 2005 still have not received support for livelihood restoration. Furthermore, new farming lands were not formally assigned to each family, despite being stipulated in the policy guidelines.[9]

These problems highlighted the need for an adjustment in the resettlement and compensation policy. The government then issued Decision 02/2007/QĐ-TTg (amending Decision No. 459/QĐ-TTg) that resulted in two key changes: extension of the application of the compensation policy to host communities and financial support to resettled families for livelihood recovery. Decision No. 02/2007/QĐ-TTg[10] addresses new issues such as monitoring of implementation, particularly with regards to the transparent management of costs and benefits of resettled people and the host communities. The total investment for the Son La resettlement programme is the highest ever in Vietnam. The estimated amount is 700 million Vietnamese dong (VND) (US$46,000) per household, which includes compensation, assistance for relocation, restoration livelihood, agricultural production support and infrastructure construction. The majority of this money has gone to infrastructure construction.[11]

More specifically, the application of the compensation policy is extended to cover host communities whose lands are acquired and/or have to move to make way for the designated resettlement sites. These host families receive assistance for investment in running water and electricity, the same as the resettled people themselves. Financial support to resettled families for livelihood recovery includes seeds, fertilizers, pesticides, perennial trees and livestock. Each household receives credit of 7 million VND (US$500) for the household head and 3 million VND (US$200) for every additional household member.

The findings of a study by Tran and Le (2008) examining this adjustment showed that the first policy adjustment in Decision 02/2007/QĐ-TTg has brought positive results by contributing to solutions to address conflicts between host communities and resettled people that have arisen directly from resettlement issues. In addition to lessening tension between the people affected by resettlement programmes, this has helped to speed up the resettlement provision of benefits to the affected communities. However, resettled people in all surveyed sites were still not happy with the second adjustment. They expected to receive higher levels of financial support for livelihood restoration (Tran and Le, 2008).

Implementation of resettlement programmes: Variation between two provinces

Son La province has the largest number of resettled households compared to Dien Bien and Lai Chau provinces. Based on detailed plans of resettlement areas, Son La province has eight regions (districts), 62 areas (communes), 237 sites for resettlement (including three standby sites) with capacity to receive 13,100 households.[12] And by the end of 2007, the People's Committee of Son La Province had approved draft detailed plans for 60 areas, 213 sites to receive 12,024 households.

In a 2008 survey, the resettlement programme was at the height of implementation. Son La province mobilized its whole administrative and political systems to accomplish the resettlement by the end of the dry season of 2009 (up to 19 May 2009[13]). The resettlement programme identified three types of resettlement (see Box 3.1). According to the plans, from October 2007 to 19 May 2008, 4606 households were resettled in 39 resettlement areas, of which 93 sites were *di tap trung* (concentrated resettlement) and 58 were *di xen ghep* (mixed sites, where the resettled people live together with the host communities). People who moved to designated concentrated resettlement sites in different communes or districts usually received more attention from the project, while people who just moved to areas surrounding the reservoir or were mixed in with host communities encountered more difficulties and sluggishness in terms of infrastructure readiness, compensation or support for rehabilitation (Dao, 2010).

Up until 20 June 2008, the number of displaced households was 8146 out of 12,400 (66 per cent) (Lai Chau People's Committee, 2008; Son La People's Committee, 2009). In general, by June 2008 there were still a large number of people needing to be resettled before the deadline of the resettlement programme in May 2009.

In Lai Chau province, as of April 2008, profiling[14] of 3584 households out of 3805 had been completed (94 per cent). Detailed plans for 100 per cent of the resettlement sites have been approved. Compensation and assistance for 3397 out of 3805 households has been completed. Draft compensation and assistance for 2755 out of 2805 households has been approved. Several component projects were

> **BOX 3.1 TYPES OF RESETTLEMENT IN THE SON LA PROGRAMME**
>
> The Son La Dam resettlement programme identified three types of resettlement, each entailing a different end situation in the resettlement area:
>
> - *di tập trung* is a concentrated resettlement, where all resettled people are moved to a single area and restart their lives together. There are two types of *di tập trung*: rural and urban;
> - *di xen ghép* is known as a mixed resettlement, meaning that the resettled people are absorbed by a number of communities;
> - *di dân tự nguyện* is the freest form of resettlement, which allows individuals to search for their own resettlement site.

Table 3.1 *Resettlement progress in Son La and Lai Chau provinces as of May 2008*

Province	Number of already-resettled households	Number of households to be resettled	Total of resettled households
Lai Chau	2320	1485	3805
Son La	8146	4254	12,400
Total	10,466	5739	16,205

Source: From author's survey and interviews

ongoing, such as construction of inter-resettlement sites, ground leveling and water pipe installation. 2320 households (61 per cent) have resettled into new places (Lai Chau People's Committee, 2008) (see Table 3.1).

Among the resettled people researched, 71 per cent are considered those falling under the 'concentrated resettlement' programme, while 26 per cent were considered of 'mixed resettlement'. The remaining people were those in the final grouping who voluntarily chose a place to resettle, rather than moving to a designated site. Because these people searched for their own resettlement sites, the time needed to adjust to their daily lives was lessened and there was also less pressure placed on already limited available land. However, the burden on individual families was much higher, as they had to deal with all the arrangements themselves including moving, buying land for housing, finding farming land and restoring livelihoods, among others.

According to the government's resettlement policies and the Son La hydropower project's policies on compensation and resettlement, the government is responsible for compensating the affected people by providing cash, supporting house moves and distributing new farmlands, seeds and stock as well as supplying rice for at least six months, and possibly up to 24 months. However, as seen in several development projects, the compensation in cash did not help the affected re-establish and readjust to their lives after being resettled. In many cases, the compensation was not invested in businesses but was used, for example, for buying goods (TVs, motorbikes, furniture) and even drugs and alcohol (VUSTA, 2006). Additional consultation with affected groups may have assisted in ensuring that compensation was directed towards business investments.

Due to the different characteristics of Son La and Lai Chau provinces in terms of natural, social and financial conditions, there were differences in people's ability to adapt to the market-oriented economy. There were also differences in the way in which compensation and resettlement programmes were implemented at the local level. The provincial authorities made decisions based on the local socio-economic and physical characteristics as well as land availability. Resettlers in Lai Chau received more land (3500–4000 m² per person) than resettlers in Son La province (about 2500–3000 m² per person). In Lai Chau province, the Provincial Resettlement Project Management Unit is the investor for infrastructure projects related to resettlement, while in Son La province, district governments are the direct investors of the infrastructure construction for resettlement areas. In Son La province, the districts' Resettlement Units are entirely responsible for moving, compensating, and supporting rehabilitation for the resettlers, while the province's Resettlement Project Management Unit only functions as a state management agency and as an investor for bigger projects with investment over 5 billion VND (about US$330,000).[15] Thus, due to its scope, the government had to give the Son La project a special mechanism in order to implement its resettlement programmes. Application of the government's resettlement policies accordingly vary from province to province under the same project.

The preceding section has illustrated two important points regarding the institutional context in which participation is carried out in resettlement programmes. First, the policy environment is often dynamic, with adjustments made along the way in order to deal with shortcomings and unexpected problems. Second, differences in the ways local governments interpret policy produce a number of differing results; in other words, people may fare differently under the same policy because of the conditions, capacities and priorities of the local authorities.

PARTICIPATION OF THE AFFECTED PEOPLE

This section discusses the participation of affected people, drawing out three main issues that have emerged as key determinants of the outcome of resettlement in the Son La project. First, the provision of information, one of the most fundamental foundations for participation, can be characterized as top-down and ineffective in its function of empowering people. Second, the timing of opportunities to participate meant that affected people were marginalized from the decisions that most affected them, preventing meaningful participation. Finally, the results of participation varied across the different ethnic groups in the resettlement area, demonstrating both the need for locally sensitive communication processes and also deep understanding of the local decision making structures and customary practices.

Communication and mobilization: A top-down approach to participation in resettlement programmes

Delivery of information and community consultation are important steps to provide affected people with information related to displacement and resettlement. Because they are the first steps, they define the scope of participation for the entire programme. Informing the community of the dam project is the first step in consultation, which is then followed by provision of information about the project's impacts and preparation for implementation. This is especially important for the case of uplands Vietnam, where affected people live scattered in remote areas without much access to information. These activities help increase the ability of affected villagers to participate in certain stages of the project (VUSTA, 2006). Lessons learned from previous resettlement programmes enabled the Vietnamese government to pay more attention to resettlement issues caused by the Son La Dam, more so than any other project.[16] This is a significant advancement in the resettlement policy in Vietnam. However, in reality the Son La resettlement project is far from being perfect in term of compliance with policy. Compensation procedures and construction of resettlement sites have been very slow, and this has had a great impact on people's ability to settle and adjust to their new lives.

The top-down method in implementing resettlement at the pilot sites of Tan Lap, Son La Province and Si Sa Phin, Lai Chau in 2003 caused many problems in these two areas. For example, housing and farming in the new locations did not match with ethnic communities' customs (Eco-Eco, 2004). A number of resettled households, including Kho Mu and Thai people in Tan Lap, have left the sites and returned to their old villages, because they were not able to become accustomed to the climate and were unable to re-establish their lives by raising dairy cows, or planting tea – activities they were not familiar with before moving. People complained to local authorities who later requested the central decision-makers to adjust the policy. Lessons learned from these two pilot sites helped change the policies applied to the whole project. For example, people no longer had to move to ready-built concrete houses, but dismantled their houses in the old villages and brought the materials to rebuild them in the new location.

Information about the project and other problems relating to the resettlement programme was mainly disseminated to villagers through local officials at village meetings. Leaflets on the Son La hydropower project resettlement plan (which affected people's rights and obligations) as well as information on compensation were printed and delivered to affected households. Even though there were only printed versions in Vietnamese, the village meetings were held in local languages.

In Lai Chau province, up until April 2008, communication was conducted through commune organizations, administration and mass media. Pamphlets with questions and answers on resettlement were given to households and resettlement officers. However, this method was not effective for two reasons. First, most villages are scattered, so it was very difficult and expensive to visit every household. Second, the level of education of people in affected areas is relatively low. Although they can generally read and write, reading the leaflets was difficult for many ethnic minority people, mostly because written materials are not traditionally important sources of information for the local people. This suggests that it is not just the creation of information products, but more importantly the format in which the material is communicated to the communities, that determines its effectiveness.

Information disseminated through meetings with commune agencies and village leaders was the most effective way of communicating, because the meetings were held in local languages and people could always ask for clarification. Most resettlers knew where they would move to and the host communities knew where resettlers were coming from. Most of the local political system participated in this campaign. Central mass media systems have also disseminated information on the resettlement programmes of the Son La project.

In both cases information provision was in a top-down manner. However, there were differences in the communication strategies and practices. Table 3.2 summarizes some of these differences.

In both cases, the distribution of Vietnamese language written materials was a key component of the governments' strategies to reach the people. In both cases, these efforts were seen to be ineffective. In the Son La case, information materials were followed up by meetings held in local languages, which made a significant improvement in both the quantity and quality of information that was delivered to affected people. To a large degree, the Lai Chau approach lacked this key aspect of communication. The Son La case is also instructive in the way it involved a number of institutions in the provision of information, including not only project-related officers, but also local media and government organizations. The lack of orthographies for many local languages is a barrier to the effectiveness of written communication materials that should be addressed in the longer term, but there is immediate scope for much more local language verbal interaction.

Our open-ended interviews show that in reality the compensation consultation was not thorough enough. Compensation was in the end a top-down decision, and the affected people were simply recipients of the decisions. As a consequence, compensation levels were much lower than market prices in terms of land for farming and housing. Many people living along main roads or in towns did not agree with the compensation levels. For example, a number of Kinh (Viet) and Thai household businesses in Muong So, Lai Chau province entered into legal claims for redress and even staged protests. They asked for compensation at market price at the time of the move. The conflicts were finally resolved through intense negotiations between the project and the affected people.

Table 3.2 *Communication strategies and outcomes in Son La*

Province	Resettled Groups	Communication strategies	Outcomes
Son La	Thai, Kho Mu, La Ha, Khang, Kinh (Viet)	Resettlement and compensation management units were formed at the district level. Officers of these units coordinated with commune authorities and village management boards to organize at least four meetings in each village. General information materials produced in Vietnamese language and disseminated at village meetings through village leaders. Village meetings held in local languages. Detailed information on rights and responsibilities, resettlement programme and compensation produced in Vietnamese language leaflets and delivered to individual households. Local mass media and political organizations involved in resettlement.	Increased effectiveness in persuading people to move. Increased methods for communication with villagers. From the leaflets in Vietnamese, people gained basic knowledge as to why they were told to move. But they still were not clear about a detailed moving plan until they actually had to move. Through numerous meetings and leaflets disseminated people obtained information on compensation policies. People increased their understanding about the whole resettlement programme as the information became available through the village loudspeaker system and village political organizations. They also knew where to move to and sometimes reacted if they did not agree with the plan. In the end people were displaced according to the plan, but a number of resettled families still could not establish their livelihoods after three years of being resettled and some returned to their old villages for farming.
Lai Chau	Thai, Kho Mu Khang, Ziao Mien, Kinh, La Ha	A Resettlement and Compensation Management Board was formed at the provincial level. This board directly deals with all the work (from communication to implementation and compensation). There is no Resettlement Unit at district level. Communication through commune organizations and mass media. Printed materials in Vietnamese languages disseminated. In some cases, resettlement officers wanted to finish the work as soon as possible, so they made the move seem more attractive to persuade people to move faster.	Less effective as officers of the Resettlement Board at the provincial level went to the village to persuade people. They were not very successful in communicating with villagers. Printed materials did not reach all households; official communication was hasty and lacked follow-up. Many resettled people were displeased with their compensation; some even wanted to fight with certain resettlement officers.

Because most affected people are farmers, they wanted to receive compensation in the form of land (to build houses and cultivate). However, due to a large number of resettled people, the amount of land allocated was very limited. Meetings with the representatives of households relating to the compensation prices of farming and housing lands were treated as a formality, rather than given serious attention. A retiree in Cho village said 'it seems that the officials in charge of such meetings just wanted to quickly finish their work'. Only the meetings about housing locations in resettlement sites took longer as people wanted to discuss the specific locations of their houses. Although in some localities people were gathered to be consulted about land for farming, the results were buried in formal reports to provincial and district authorities. There was never any response, but it was believed to be mainly because of serious land shortages in resettlement sites. In addition, other needs such as water and electricity in some of the resettlement sites were not available when people moved in, or compensation was very slow, making people extremely frustrated. The provincial and districts authorities were unable to solve these problems by the time people complained, so they avoided confronting them.

A Thai farmer in Cho village (Lai Chau Province) said: 'It is true that we have been informed about the resettlement policies and compensation, but in fact attendees of such meetings were different, the wife comes this time, her husband or son comes the next time, and they often do not inform each other although they are one family.'[17] As up to seven years passed, people also forgot what they had heard about the resettlement programme as they waited to be resettled. Because they waited such a long time, local people became less comfortable in investing in their houses, for example, for repairs. Having learned that their village would have to move, many households prepared timber for their new houses, but as time passed the timber began to degrade.

The implementation of social security policies requires more attention for more vulnerable groups such as ethnic minorities, landless people, women, the old and the poor, among others. However, these groups often do not receive attention from the project staff. Generally, women-headed and poor households are the most negatively affected because they are hardly able to raise their voice. The cases of Pu Nhuong and Phieng Bung in the following section provide examples of this situation.

Ethnic minority groups with small populations also suffered more during resettlement. For example, a Giay village (a host community) with 19 households in Then Cho only received about one-sixth of the market value for their land lost. According to the headman, they did send redress to the project, but the answer was their land was rotated farming land and did not have a high value. And the price was set at the time when decision on land recovery was made, not when they actually moved. They had to accept the government's set price despite it not reflecting the market price. Therefore, there were many cases of complaints, and in some villages people did not move as planned (for example Muong So, Phong Tho town, or Muong Khieng resettlement site). They delayed the resettlement even though they eventually had to move to the designated sites.

Participation in different stages of the process

As mentioned above, the Son La Dam has been constructed in order to meet the country's increasing demand for energy. Before the government made its final decision, a number of studies with regards to technology and impact assessments on social, cultural, ethnic, livelihoods and environmental conditions were conducted by professional institutions such as the Institute of Ethnology and the National Institute for Agricultural Planning and Projection (NIAPP), among others. The National Assembly made the final decision on the project. As decisions were made from the top, there was no opportunity for affected communities to participate during this stage of the project (VUSTA, 2006).

In the planning stage for the project, a master plan was prepared. Additional surveys on socio-economic and cultural conditions of affected people were conducted, but at this stage, generally, affected people were only able to participate in the socio-economic surveys.

In the technical design stage (including the design of the dam as well as the resettlement sites), affected people had a chance to participate, especially with regard to places to which they were going to move. Moving plans were determined through surveys about people's work, economic situation and land requirements. Only at this stage could people express their opinions with regard to resettlement. Officials were sent to villages to consult with the villages' headmen and the elderly about their choices of resettlement sites. They then organized meetings in which villagers contributed their ideas for infrastructure such as how roads would be built, and where schools and clinics should be located. The meetings were held in local languages and interpreters were provided if project officials were in attendance. In the pilot resettlement sites, there was no consultation regarding these issues.

In the implementation stage, management skills were deemed necessary for building infrastructure such as electricity, water supply, houses, healthcare centres and schools. Districts were responsible for undertaking this work and people only built their houses. However, in the first set of displaced villages, such as the pilots in Tan Lap and Si Sa Phin, people did not have an opportunity to build their houses, because houses and other infrastructure were already built for them when they moved in.

The final stage of resettlement involved completing and maintaining infrastructure. As planned, infrastructure for new villages is guaranteed for three years by the government and then responsibility is turned over to local people. In fact, there are many problems with infrastructure in the resettlement villages due to difficult weather conditions in the northwest's mountainous areas. After only a couple of years, roads and water pipes in many villages were seriously damaged by rain, floods and landslides. Repairs have been very slow. For example, the author's interviews in 2008 show that after a three-year guarantee, there is no longer water supply in Pu Nhuong resettlement site. The road to the village was totally degraded. According to the head of the village, more than two years have passed since the water supplies and road were damaged and although many project officers visited the village a number of times to take measurements, no progress was made on repairing the much-needed infrastructure. On the other hand, local authorities cannot mobilize funding for maintenance from the poor villagers, who struggle to make a living once the government subsidies end.

Ethnicity and participation

Affected people living in areas to be submerged under the Son La hydropower reservoir are mainly minority ethnic groups, as mentioned previously. Many of these communities have had little access to formal education and have been marginalized from decision-making processes. Literacy in Vietnamese is limited, especially among women. This is because many women living in remote villages dropped out of school when they finished grades two or three (usually around the age of nine or ten) and stayed home to help with housework and farming. These factors make it difficult for ethnic minority communities, especially women, to access information about the project, its impacts and the details of the resettlement programme. In many cases, these communities are not aware of their rights within the project, or even within Vietnamese society at large. With low levels of Vietnamese language capacity, the standard approaches to information provision and consultation are ineffective. The knowledge and value systems of ethnic minorities often vary, not to mention that they are notably different from the Kinh.[18] This means that not only is there a need for communication that is appropriate for a non-Kinh audience, but even more, there is a need to recognize differences within the diverse upland population.

The government has tried various methods of communication with local people on these issues. The first step towards establishing good communication and good participation in a multi-ethnic situation is the establishment of trust, often through the mediation of the customary leadership of a community. However, in reality this is a very complicated set of dynamics, and it takes time for people to re-establish and adjust to their new lives. The more they are able to participate in making decisions related to resettlement, the faster they will be able to re-establish their livelihoods.

There is a high level of diversity within upland societies. In the resettlement sites of Huoi Luong (Lai Chau Province) and Pu Nhuong (Son La Province), the participation of Giay and La Ha people is clearly different from that of the ethnic Thai. In Pu Nhuong, despite the presence of high-ranking officials on the village management board, they rarely gave their opinions when discussing securing people's livelihoods and village production, except when asked. This is probably because they are in the minority compared with the Thai people, but also because their living standards are generally lower. La Ha people in general have lower levels of education compared to the Thai. There are no La Ha people working in the commune authority. Thai families in Pu Nhuong were somehow given more and better farming land than La Ha families even though the headman carried out a lottery system for villagers to pick their own farming plots. As a consequence, most of the La Ha families returned to their old villages for farming. In Huoi Luong, although 19 Giay families share their farming land with Thai resettlers from Chan Nua with the same compensation level, their living standard is considered lower than that of the Thai people. When taking part in detailed resettlement planning, the Giay ethnic group merely agreed to the plan without any objection. Perhaps, they trusted the headman and resettlement staff, but it is also likely that they were not able to participate meaningfully in the consultation because of language, limited education or ineffective consultation processes. Since the resettlement, the Giay people are the poorest in Huoi Luong even though they were the ones who had to share their land for the resettled.

Why does a difference in participation by ethnicity remain? According to local authorities, it is because of lower levels of education, reluctance to communicate, relatively limited ability to adapt to changes compared to Thai people[19] and unplanned expenditures. Management of income and expenditure is not well organized and after harvesting, money is spent on items of interest as opposed to items of need. The concept of savings is not well established and thus during difficult times excess funds are not available. Thai people tend to save more. Although the causality of factors and outcomes is not clear, there were several cases where it was obvious that resettlement officers did not make necessary efforts to ensure that communication was effective when working with ethnic minority communities.

Besides the above-mentioned differences, there remains gender differentiation and internal conflicts inside many ethnic communities. The resettlement areas of Pu Nhuong or Phieng Bung are examples. Women are not selected as a head of the village or in the village's management board. The only position for women in a village is as a representative of the village's Women's Union. People with high economic capacity, good social networks and who obtained higher education are commonly voted to be in the Village Management Board or head of socio-political organizations (for example the Women's, Youth or Farmer Associations etc.). In general, they are leaders in many activities in the villages. Sometimes, they make decisions on behalf of the village without consulting properly with their community. In this case, village meetings are used simply to legitimize their unilateral decisions by collective agreement. Contrasting opinions are rarely discussed or adopted. Villages and communes have their own informal traditional social structures, which have been fractured by the resettlement process, affecting active participation of ethnic groups. Even where resettlers and hosts are from the same ethnic group, it is very common that resettlers are less socially active in the community.

The main problem is that although there was an improvement in people's participation in the Son La resettlement project in comparison with others, the quality and level of participation is a critical issue. Efforts must be made from two sides: from the state to create opportunities for meaningful participation; and from civil society and affected people to take advantage of the government's improvement in policies.

Conclusion: Information, Consultation, Cultural Sensitivity and Empowerment

There has been a remarkable improvement in the resettlement policies of Vietnam over the last couple of decades, reflecting the importance and effects of community participation in resettlement projects. People gradually recognized that top-down decisions, which were not compatible with their situation, negatively affected their lives both in the short and long term. However, there remain significant barriers to meaningful participation of local communities in the key areas of decision-making ranging from ineffective government policies and resettlement officers' attitudes to local people's ability to participate depending on their ethnic norms and practices and geographical locations.

The implementation of the Son La resettlement programme has not led to the desired equitable socio-economic development outcomes as expected. Neither has it been successful in compensating the affected communities for loss of livelihoods because of the dam construction. Nonetheless, the Son La Dam resettlement process has progressed in the involvement of affected people's participation during project implementation relative to other projects. In fact, even though the involvement of people in the resettlement process of Son La hydropower is still very limited, it is considered to be much better than other dam-induced resettlement projects such as the Thac Ba, Hoa Binh, Tuyen Quang and A Vuong dams elsewhere in Vietnam. Affected people are expected to continue their engagement with developers and project planners to bring about adjustments to the implementation of the programme, especially in the context of the growing voice by civil society groups in the last few years such as the Vietnam Union of Science and Technology Associations (VUSTA) and Vietnam Rivers Network (VRN), among others.

This chapter has examined the process of implementation on the ground, and found that the participation of affected people and communities in the resettlement of the Son La hydropower project is by no means uniform or straightforward. Cultural differences affect political and legal institutions in the provinces. However, there is a need for more detailed studies to understand the role of culture in consultation processes such as those discussed here. It is clear that ethnic minority communities face a range of hurdles in achieving meaningful participation, including language, difficulties with bureaucratic logic and communication and the limited scope of rights to engage in decision-making processes.

This chapter also shows that participation must be included in the planning stages, especially when the details of resettlement areas and support services are discussed. Transparent monitoring of the plans is crucial, and can be improved if local people have been involved in the process from the planning stages. In other words, accountability mechanisms focused at the local level of implementation can be created through involvement of affected peoples.

The chapter has thus highlighted two main lessons from the Son La experience. First, participation is not carried out in a vacuum. The dynamics of policy formulation and reformulation at the central level, coupled with differences in implementation strategy among local governments, demonstrated this. Second, there is a complex mix of interrelated factors that influence how local people participated in the resettlement scheme. Capacity building and empowerment of the local people must go hand in hand. At the same time, sensitivity, participatory skills and accountability of the local authorities must be considered, particularly in a situation of extreme ethnic and socio-economic diversity. These findings have implications for the degree of influence that the global debate on water governance has on the ground in any country. This analysis also suggests that one practical challenge for improving the governance of water resources is pushing opportunities for participation into the higher levels of decision-making, where the scope of critical issues such as resettlement programmes are determined.

Dam-related involuntary resettlement is only a part of bigger issues of development, which determine water governance and river basin management in Vietnam. Lessons learned from participation practices in resettlement programmes may guide government policies in dealing with these issues more effectively in the future. If genuinely participatory processes can be achieved broadly in specific projects, it is hoped that the more important decisions about whether or not to build a dam will be opened up to a larger range of voices, especially those of the local communities themselves.

Notes

1 Three articles in UNDRIP that explicitly require FPIC include: 'No relocation shall take place without the free, prior and informed consent of the indigenous peoples concerned and after agreement on just and fair compensation…' (UNDRIP, 2007, article 10), 'before adopting and implementing legislative or administrative measures that may affect them' (UNDRIP, 2007, article 19); 'and prior to the approval of any project affecting their lands or territories and other resources, particularly in connection with the development, utilization or exploitation of mineral, water or other resources' (UNDRIP, 2007, article 32).
2 Very recently, the UN-REDD (Reduced Emissions from Deforestation and Forest Degradation) programme held consultations with Indigenous Peoples from Asia and the Pacific to look at how to apply the principles of FPIC to the REDD+ Readiness process. A workshop was held in Vietnam and members of national UN-REDD programmes are underway to implement national-led activities (UN-REDD Programme, 2010). Whilst not specific to hydropower and resettlement, this work signals a step in the right direction and could be applied in the future.

3   The old saying that is often used to describe this situation and to condemn the investors (state-owned companies) is 'Đem con bỏ chợ' (literal translation is 'to hand over one's responsibility to others without knowing who they are').
4   In 1998, the ADB funded a project that assessed resettlement policies of seven borrower countries, including Vietnam, in order to encourage these governments to improve resettlement policies (Pham and Lam, 2000).
5   The Land Law 1993 specifies the rights and obligations of people who have been assigned or leased land (called land users). It gives them the right to change, transfer, lease or mortgage land. Furthermore, they are entitled to compensation for any land loss. The Land Law was an important landmark for improvements in resettlement policies and in planning processes for development projects.
6   In this chapter we discuss resettlement issues in the Son La and Lai Chau provinces only.
7   The resettlement was divided into two levels: 145 metre level and 218 metre level. The resettlement for the 145 metre level was completed by the end of 2005 and was only for the Son La province. The 218 metre level applied for the remaining resettled communities that are in the three provinces.
8   Program 135 is the Prime Minister's decision No. 135/1998/QĐ-TTg to support socio-economic development for ethnic minority people in difficult mountainous regions.
9   Author's survey in 2008.
10 Decision No. 02/2007/QĐ-TTg by the prime minister on the regulations of compensation, assistance and resettlement of the Son La Hydropower project. The supporting level is lower than that in Decision No. 01.
11 Author's interview with resettlement officers in Son La and Lai Chau in 2008
12 Of which 59 areas and 223 sites for rural resettlement with capacity to receive 10,574 households; three areas and 14 sites for urban resettlement with capacity to receive 1661 households; resettlement with existing local people to 60 villages of 23 communes with capacity to receive 755 households and 110 voluntary resettlement households.
13 Ho Chi Minh's 120th birthday.
14 The project needs to have a complete profile for each household including information such as amount of lost land, damaged properties, all kinds of compensation received after moving, etc.
15 By the time of our field study, resettlement was not implemented in Dien Bien province. This was because the resettlement sites in the province were at the highest level of the reservoir (200–218 metres) and were not considered urgent resettlement sites.
16 Government Decree No. 29/1998/NĐCP and Prime Minister's Decision No. 30/2005/TTg require that interests as well as obligations of affected people should be considered during the process of making decisions.
17 Author's interviews in 2008.
18 Our studies and observations reveal that Kinh women are usually much more active in raising their voice in resettlement-related matters. Ethnic minority women including Thai, Kho Mu, La Ha and Giay barely participate in village meetings. That may be due to their shyness or limitation in social activities. Unlike Kinh women

(who usually have junior or high school degrees) many ethnic minority women only finished grade two, or did not go to school at all. They don't usually deal with issues outside of the home and most of their time is spent taking care of children and doing housework. There are, however, some exceptions where people have had opportunities to be trained at district or province level, or in special schools for ethnic minority children.

19 Author's interviews with local authorities.

REFERENCES

ADB (1992) 'Manual on resettlement: Guideline for implementation', Hanoi Publisher, Hanoi, Vietnam

Cam, T. and Ngo, Duc Thinh (1999) *Customary Law of the Thai in Vietnam*, Nationalities Culture Publishing House, Hanoi

Cariño, J. and Colchester, M. (2010) 'From dams to development justice: Progress with "free, prior and informed consent" since the World Commission on Dams', *Water Alternatives*, vol 3, no 2, pp423–437

Cernea, M.M. and McDowell, C. (eds) (2000) 'Risks and reconstruction: Experiences of resettlers and refugees', World Bank, Washington DC

Cook, C. (ed.) (1993) *Involuntary Resettlement in Africa*, World Bank, Washington DC

Cooke, B. and Kothari, U. (2001) *Participation: The New Tyranny?* Zed Books, London

Dang, N.A. (2007) 'Resettlement for hydropower projects in Vietnam', *Communist Party Review*, 1 August 2007, Hanoi

Dao, T.H. (2006) 'The study on the livelihood and environment at resettlement area of KleiKrong Hydropower Project', Kon Tum province, presentation to Vietnam Rivers Network annual meeting in Hanoi, 5 December 2006

Dao, N. (2010) 'Dam development in Vietnam: The evolution of dam-induced resettlement policy', *Water Alternative*, vol 3, no 2, pp324–340

Diep, D.H. (1995) *The Transformation of Ethnic Communities. The Impact of Hoa Binh Reservoir*, Social Sciences Publishing House, Hanoi

Doan, B. and Nguyen, D.A. (2006) 'Assessment on life quality of resettlers in Ta Trach Hydropower project, Thua Thien Hue province', presentation to Vietnam River Network's annual meeting in Hanoi, 5 December 2006

Duong, A.T. (2009) 'Assessment of consultation process with affected people in Bac Ha hydropower project, Lao Cai province', presentation to Vietnam Rivers Network annual meeting Hai Phong city, 27–29 November 2009

Eco-Eco (Institute of Ecological Economy) (2004) 'Assessment on living conditions and potential economic development for people at resettlement sites of Son La Hydropower', Tan Lap site, October 2004

Government of Vietnam (1959) Decree No. 151/ 1959/NĐ-TTg dated 14 April 1959 of The Prime Minister of Vietnam on Degine advance provisionally when State Recovers Land

Government of Vietnam (1993) Land Law

Government of Vietnam (2004a) Decision No. 459/QĐ-TTg of Prime Minister on compensation and resettlement in Son La Hydropower Project dated 12 May 2004

Government of Vietnam (2004b) Decree No. 197/2004/NĐ-CP dated 13 December 2004 of the Government of Vietnam on Compensation, Assistance and Resettlement when State Recovers Land

Government of Vietnam (2004c) Decision No. 196/2004/QĐ-TTg by the Prime Minister on regulations of compensation, displacement and resettlement

Hirsch, P. (2001) 'Globalisation, regionatlisation and local voices: The Asian Development Bank and re-scaled politics of environment in the Mekong Region', *Singapore Journal of Tropical Geography*, Department of Geography, National University of Singapore and Blackwell Publishers Ltd, pp22–23

Hoang, L.A. (2006) 'Assessing life quality and potential for rehabilitation of the resettlers in A Vuong hydropower project', presentation to Vietnam Rivers Network annual meeting in Hanoi, 5 December 2006

Hoang, N.G. and Vo, H.C. (2006) 'Study on socio-economic environment of resettlement area of Thac Ba Hydropower Project, Yen Bai, 32 years after the dam's construction', presentation to Vietnam River Network annual meeting in Hanoi, 5 December 2006

Institute of Geography (2005) 'Report on ethnology and socio-economic issues of ethnic minorities groups affected by Son La project', Hanoi, Vietnam

Lai Chau People's Committee (2008) 'Report on the implementation of compensation in the Son La hydropower's resettlement in Lai Châu province by April 2008, Report No.19/BC-BQLDA (dated 23 April 2008)

Le, K.S. and Nguyen, T.T.P. (2006) 'Resettlement of Son La Hydropower Project from a socio-economic perspective', component report for VUSTA, Hanoi

Middleton, C., Garcia, J. and Foran, T. (2009) 'Old and new hydropower players in the Mekong Region: Agendas and Strategies', in F. Molle, T. Foran and M. Käkönen (eds) *Contested Waterscapes: Hydropower, Livelihoods and Governance*, Earthscan, London

NIAPP (2004) 'Report on resettlement planning of Son La Hydropower Projects', National Institute of Agricultural Planning and Projection, Hanoi

Pham, T.M.H. (1997) *Basic Differences Between International Organizations and Vietnamese Laws Regarding Development-induced Resettlement Policies*, Hanoi Publisher, Hanoi, Vietnam

Pham, T.M.H and Lam, M.L. (2000) *Resettlement in Development Project: Policy and Reality*, Social Sciences Publishing House, Hanoi

Pham, X.N. (ed.) (1995) *Rural Development*, Social Sciences Publishing House, Hanoi

Ribot, J. (1998) 'Theorizing access', *Development and Change*, vol 29, pp307–341

Singh, S. (2009) 'World Bank-directed development? Negotiating participation in the Nam Theun 2 Hydropower Project in Laos', *Development and Change*, vol 40, no 3, pp487–507

Son La People's Committee (2009) Son La Resettlement and Rehabilitation Board, Report No. 81, September 18, regarding implementing results of the resettlement plan in 2009

Son La Province's People Committee (2006) Son La Hydropower Plant and its Resettlement, Son La Province

Tran, V.H. and Le, K.S. (2008) 'Study on impacts of resettlement of Son La Hydropower Project', Center for Water Resources Conservation and Development (WARECOD), Hanoi

UNDRIP (2007) 'UN Declaration on the Rights of Indigenous Peoples', New York, United Nations, available at www.un.org/esa/socdev/unpfii/en/drip.html (last accessed 10 June 2010)

UN-REDD Programme (2010) 'UN-REDD hosts FPIC Workshop in Viet Nam', Issue #10, July, available at www.un-redd.org/Newsletter10/UNREDD_FPIC_Workshop/tabid/4860/language/en-US/Default.aspx (last accessed 19 July 2010)

Vo, V.H. (2006) 'Study on environmental and social situation at resettled areas of Ban Ve Hyropower Project, Nghe An', presentation to Vietnam Rivers Network annual meeting in Hanoi, 5 December 2006

VUSTA (2006) 'Study on impacts of the Son La Hydropower Resettlement', Vietnam Union of Science and Technology Associations, Hanoi

Part II

Social Differences and Access

4

Rights and Rites: Local Strategies to Manage Competition for Watershed Resources in Northern Thailand

Nathan Badenoch and Prasit Leepreecha

INTRODUCTION

The 1980s marked the beginning of the end for the opium economy that had dominated the upland landscapes of northern Thailand for several generations. Within 10 years, the brightly colored poppy flowers had been replaced by green fields of cabbage and shallot. The daily concerns of the upland farmers were not limited only to availability of land, but were increasingly about access to the sources of water for irrigation. Gravity-fed irrigation systems allow farmers to crop in the dry season, placing new pressures on the streams that originate in the upper tributary forest areas. This upland agricultural economy initially brought about by a mixture of government policy, pressure from lowland society and local technological innovation is now characterized mainly by competition for natural resources.

The successful replacement of poppy with other commercial crops has meant changes to the ecosystems that support the people of this area. While the nature, cause and extent of these changes are hotly debated, the resource competition is viewed as having many dimensions – inter-village, inter-ethnic, upstream-downstream, minority-majority – within the larger context of upland watershed management that has come to dominate the environment-development debate in Thai society. This debate is in large part about the legitimacy of upland people – more often than not ethnic minority communities – in upper watershed areas. Facing not only the daily difficulties of resource competition in their livelihood activities, upland farmers have long been disenfranchised by the natural resources policies, as well as the development interventions more generally, of the Thai state (Hirsch, 1990). However, upland people are responding to these multifaceted challenges through a range of approaches.

This chapter explores the response of the Hmong people, an ethnic minority group that has been at the centre of the agricultural and economic transition of upland landscapes, to these environmental, social and political pressures. When they first arrived from China in the late 1800s, they were a highly mobile group of people supporting themselves by cultivating upland rice and poppy. In response to state efforts to halt poppy cultivation in the uplands beginning in earnest in the 1980s, Hmong communities began replacing opium poppy with cash crops and fruit trees, in many cases leading other upland communities, as well. Previously reviled for their involvement in the opium economy, the Hmong now face accusations of destroying forest and water resources through the chemical-intensive agricultural practices that were recommended by those crop replacement efforts. The Hmong have drawn on their economic, social and cultural resources to respond to local resource tensions, while at the same time asserting the legitimacy of their livelihood practices within watershed governance discourses.

In the search to not only understand, but also contribute to the possible options for dealing with upland resource challenges, community-based natural resources management has been proposed as an alternative paradigm to more top-down approaches preferred by the state. The body of research on common property resources has provided many examples of how communities organize access and manage shared resources, at many scales and in different socio-ecological settings.[1] Much of this research has focused its attention on the creation of institutions for resource management at the local level, based on local knowledge, customary practice and adaptive capacities.

In Thailand, there have been two main contexts framing research and advocacy on the potential for and experiences of community-based, or at least locally oriented, management of watersheds.[2] First, the watershed has been conceived of as a system of interlinked common property resources, which require the right institutions to regulate users as they compete for resources. Some have accordingly pushed for a *rights-based* approach to community-based natural resources. Second, watersheds have been understood as a place where local people struggle against the state in a clash of cultural symbols and knowledge systems. Examination of the importance of identity and the adaptation of traditional practices stress a more *rites-based* approach to understanding local resource dynamics. Throughout this chapter, we use 'rights-based' to refer to formalized and often legalistic approaches to control over resources. These approaches are expressed in the language of the state, and are concerned with tenure and property as provided for in a national legal framework. The term 'rites-based', on the other hand, is motivated by an alternative approach to resource access that is framed in the cultural norms and knowledge of local people, and expressed in the symbolism of ethnic identity through ritual practices.

The uplands, where this story of rites and rights over water resources is set, form an ecologically important component of the Mekong region. Socially, the uplands are characterized by extreme ethno-linguistic and cultural diversity, much of which crosses national boundaries. In recent years, it has become clear that while the viability of upland ecosystems is coming under increasingly high stress from rapid economic development, upland people share in less of the benefits from the region's economic growth, and ethnic groups are facing both political marginalization and ecological risks (Cornford and Matthews, 2007). Indigenous people of the Mekong region are subjected to impacts from economic and political forces and, in turn, devise strategic responses (Leepreecha et al, 2008). This chapter draws on the authors' research into the local dynamics of Hmong society in northern Thailand to highlight how strategies for coping with social change at the interface of local, national and global arenas are devised. Mechanisms for negotiating access to increasingly contested and scarce upland water resources are key to these strategies.

CONTESTING WATERSHEDS: WATERSHED DISCOURSE AND THE HMONG RESPONSE

Loss of forest and the associated perceived environmental impacts on lowland society have catalysed the creation of a strong battery of conservation policies in Thailand. These policies – particularly the creation of protected forest areas and the control of upland use through watershed classification – are couched in public perceptions about the fundamental incompatibility between communities and forests in critical watershed areas (Daniel and Lebel, 2006; Usher, 2009). Within this context, upland minorities have been portrayed as destroyers of the forest, and the Hmong have featured prominently in this portrayal. The Royal Forest Department (RFD) has been the main force in the formulation and implementation of these state policies (see Chapter 5, this volume). Establishing Watershed Units in localities, the state has sought to regulate use of resources in areas deemed ecologically sensitive by bringing both land and communities under the legal jurisdiction of the national forestry policy framework.

In northern Thailand, where policy has severely limited legal rights over upland forest resources, local rights over watershed forest resources have been constructed through the development of community forestry and watershed networks since the early 1990s (Na Ayuthaya, 1996). The process of rights construction, based on the understanding of forests as cultural systems, continues today, but within the context of intensified upland agricultural production. For these communities, this is also a process of establishing eco-political spaces, where bundles of rights and local histories are used to give new meaning to natural resources (Peluso, 2005). An upland watershed discourse was born of the friction between the centralized, state-led vision of watershed conservation and the locally generated conception of community management of watershed resources (see Chapter 5, this volume). It is within this discourse that the Hmong have defined their approaches to dealing with competition for watershed resources.

Conservation and conflicts over rights to resources

The establishment and expansion of protected areas and the watershed classification – both of which are legal barriers to upland farmers' legal tenure over land and other natural resources – have been the two central policies in the state effort to assert protection over northern Thailand's watershed forest areas. The result of these environmental policies is that upland communities face disenfranchisement from the natural resources upon which their livelihoods depend and exclusion from these debates that affect their future. Thus, watershed policy in Thailand has focused on the communities – predominantly ethnic minority peoples – in the uplands as responsible for the problems of dry-season water shortages faced by the lowlands (Usher, 2009). The watershed discourse in northern Thailand has thus turned lowland water shortages into a political tool involving ethnicity, by which the predominantly ethnic Thai lowlanders claim rights to decide the fate of upland resources (Laungaramsri, 2001).

In response to this force, a counter-movement has arisen placing local ecological knowledge and practices at the core of a 'human-forest harmony' discourse. The Karen people, another upland ethnic group, have been at the centre of this development. Western and Thai scholars have been joined by charismatic and articulate Karen leaders to assert the heritage of forest protection and ecological sustainability that exists within the Karen cultural traditions. While significant progress has been made in communicating the Karen perspective on these development dilemmas, other scholars have argued that the power of this movement has in fact resulted in a constriction of political space for the Karen, who have had their legitimacy and aspirations as upland agriculturalists circumscribed by this constructed identity of 'upland forest guardian' (Walker, 2001).

Environmental narratives that oversimplify the landscape interactions are at the heart of many of the watershed problems of northern Thailand (Forsyth and Walker, 2008). In this political debate, polarized positions posing ethnic minority groups as either the destroyer or protector of the forests has emerged. Both are affected by selective manipulation of information and data. The Hmong face narratives that brand them as not only destroyers of the forest, but as having overstepped the line of acceptability by going enthusiastically into commercial agriculture, which has always been considered the economic territory of the lowlands. Originally it was the environmental threat of shifting cultivation, now it is that of commercialized agriculture that defines the public understanding of the Hmong. In either case, the implications are the same: rights to resource access are limited in the name of a greater public interest, and the legitimacy of uplanders' claims over resources are denied. Walker (2003) asserts that 'responses to these denials of legitimacy are often framed in similarly ethnic terms as they promote traditions of indigenous resource management as a basis for local identification and political mobilization'.

In 1998, a high-profile upstream-downstream conflict in Chiang Mai province brought watershed tensions to a head. Believing that Hmong lychee orchards in an upstream area of Chom Thong were the reason for water shortages, lowland farmers blocked the road to the Hmong village. Lowland conservationists built fences (sometimes painted in the colours of the Thai flag) around forest areas to prevent Hmong farmers from entering. The Chom Thong Conservation Club and the Thammanat Foundation, two environmental NGOs with some well-known Bangkok-based conservationists in their ranks, led this opposition against upland Hmong farmers. Two years later, in Nan province, the road to a Hmong village was blocked by lowland farmers who entered Hmong farms and cut down lychee trees and burned houses in protest at the Hmong farmers' upland agricultural practices. The Royal Forest Department has in most cases been a passive observer in these conflicts. But both the conservationists and the RFD have used questionable scientific information purporting to show the adverse impacts of upland farming on lowland water availability (Wittayapak, 2008).

The Hmong responses presented below illustrate how a mix of rites and rights approaches are employed in efforts to establish new management institutions at different levels of governance. First, at the level of broad social discourse in northern Thailand, an environmental movement based on the redefinition of cultural symbols arose out of the highly publicized conflict between lowland and Hmong farmers. Second, at the local level, the same cultural symbols and practices were used in a Hmong community to deal with competition for resources in a period of rapid economic and landscape change. At the same time, more formalized institutions, focused on water management, were created to deal with competition between the Hmong and their Karen neighbors. Figure 4.1 shows the nested issues explored in this analysis.

The following sections will first discuss the role of the *ntoo xeeb* spirit tree in the activities of the Hmong Environment Networking Group in northern Thailand. Next, we turn to the villagers of Ban Phui Nua, a Hmong village in Chiang Mai province, where the *ntoo xeeb* was a central part in that community's efforts to manage resources collectively during a period of rapid economic change. Finally, the water management institutions developed between the Hmong and their neighbours in the Huai Sai Khao sub-catchment, under conditions of water scarcity, are presented. In the analysis, the clear distinction between rites and rights is lost, as the approaches become mutually supportive of the Hmong response to resource competition. These dynamics are then discussed within a context of multi-level governance, with a focus on the concept of resource access.

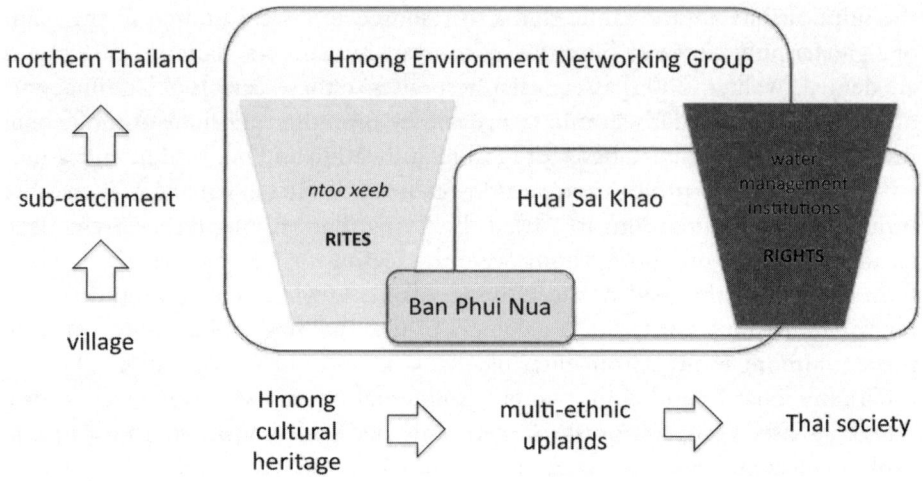

Figure 4.1 *Rites and rights in watershed management*

Source: author

Watershed rites: *Ntoo xeeb* and the birth of a Hmong environmental movement

The 'Hmong response' to changes in the natural landscapes and policy environment of Thailand has been of some interest since the 1970s, when the forces of forest conservation and poppy eradication became a local reality for upland farmers (Cooper, 1984). More recently, in the face of intensifying pressure from state efforts to manage the country's watersheds, the Hmong of northern Thailand undertook a remarkable project of 'cultural redefinition' in order to create the cultural and political space needed to assert themselves as legitimate actors in upland watersheds (Leepreecha, 2004). This social project involved three main components: the establishment of a network to work on environmental issues, the reformulation of traditional cultural symbols of human–environment interaction and the commitment to addressing local resource conflicts. The Hmong were participating in a larger movement to assert the environmental awareness and competence of upland minority groups. Until this point, the Karen had gained recognition as protectors of upland forests, as they were successful in articulating publicly how their traditional livelihoods paid respect to the natural environment and were based on certain indigenous concepts of sustainability.

The Hmong Environment Networking Group was established in 1997 in Chiang Mai and Khamphaeng Phet provinces in northern Thailand. The network quickly grew from its original mandate of promoting local forest conservation and traditional knowledge within the Hmong community to a broad-based and high-profile campaign to redefine their position in the upland watershed areas.

The network launched this effort with a ceremony for the *ntoo xeeb* of the village of Ban Mae Sa Mai, in Chiang Mai province. *Ntoo xeeb* is a tree in the landscape of a Hmong village in which the local area spirit (*thwv tim*) resides. The *thwv tim* is the greatest of the local area spirits, and represents the supernatural force that governs a mountain or watershed (Leepreecha, 2004). In the past, local practices in making offerings at the *ntoo xeeb* differed across Hmong society, but in general villagers were prohibited from cutting trees in areas adjacent to the *ntoo xeeb*.

The Ban Mae Sa Mai *ntoo xeeb*,³ a prominent but solitary tree, had been an important part of the local Hmong ritual landscape since 1965, mediating the human–natural relationship of the village as it grew. The role of the *ntoo xeeb* was transformed in 1985, when community leaders established a specific area with clear boundaries around the *ntoo xeeb*. This community conservation area was supported by regulations, and gradually expanded by the village leadership. In 1998, The Hmong Environment Networking Group, along with the IMPECT Association (an NGO of upland ethnic peoples, based in Chiang Mai, Thailand), helped Ban Mae Sa Mai to organize a public *ntoo xeeb* ceremony. Government officials, media, academics and community leaders from upland and lowland villages were invited to take part in the ceremony, including a forum on community rights and decentralized management of natural resources. This high profile event can be understood within the larger context of Hmong cultural reconstruction of ethnic identity that broadcasts its message to the national and global communities through audio-visual media, publications and the internet (Leepreecha, 2008).

In addition to *ntoo xeeb* the Hmong Environmental Networking Group helped villages to revive, strengthen and publicize their ritual practices and their spiritual linkages to the physical landscape. Offerings to the watershed spirit (*xyeem hauv dlej*) signified community understanding and appreciation of the relationship between forest management and water availability, which was particularly relevant in the context of intensifying competition for water. But this cultural practice, too, was given new political meaning in a symbolic statement made by the Hmong to Thai society (Wanitpradit, 2005). In the past, this was reserved for times of ecological or social crisis – such as crop failure or sickness – but was modified to represent the Hmong community's commitment to watershed management.

This assertion of traditional cultural heritage and environmental awareness was not only a reaction to the rising animosity towards Hmong and their agricultural practices. The *ntoo xeeb* ceremony was also an appeal to the 1997 Constitution of Thailand, which guaranteed legal rights to community participation in natural resources management. Drawing on their cultural heritage to claim recognition of their position in society, the Hmong organized themselves to engage in the discourse on watershed management on a scale never seen before. These efforts targeted national policy and general public perceptions, making a bold statement about the Hmong as an ethnic minority group within the Thai nation state. This assertion of their legitimacy in the upland landscape was a turning point for the Hmong in their struggle for rights to livelihood, and can be understood as part of

> **BOX 4.1 CREATING CULTURAL SPACES WITHIN WATERSHEDS**
>
> The Hmong created a culturally defined area of watershed protection forest around the originally narrower *ntoo xeeb*. The Karen of Mae Khong Kha watershed in Mae Chaem district brought about a similar transformation in their sacred tree practices as well. According to Karen traditions, the umbilical cord of a newborn child (*dei paw*) is placed in a bamboo container and attached to a tree that is chosen by the parents. The link between the child's life and the tree's life mean that community members cannot cut the tree. In the past, these spirit trees were scattered across the landscape. In the early 2000s, facing pressure from the Inthanon National Park authorities and the Royal Forestry Department, the Karen villages of Mae Khong Kha decided to delimit a broad area of forest for conservation. In order to gain broad-based support from the communities, villagers were asked to participate in a large ceremony to create a birth spirit forest (*sei dei paw*) and attach symbolic bamboo containers signifying the linkage between the life of the communities and the forest. This move was reminiscent of the popular Thai practice of ordaining trees (*buat pa*), but was couched in entirely Karen cultural terms. The activity was supported by the local watershed management network, and was seen as an effort to respond to the mainstream conservation movement in Karen terms that could be easily understood. The basic process at work is giving a new spatial definition to a ritual practice and attaching a broader set of institutions to regulate the community's interaction with the forest. (Authors' fieldwork with International Center for Research in Agroforestry, Mae Chaem District, Chiang Mai Province, 2003–2005)

a larger movement of upland people to establish themselves as legitimate managers of watershed resources. Box 4.1 on creation of cultural spaces within forests shows the intersection of Karen and Thai spiritual practice in the struggle to assert rights over forest resources.

Furthermore, the *ntoo xeeb* ceremony came to represent unity within the Hmong community, and the recognition that they needed to work together to transform the way in which they interacted with their local environment on one hand, and larger society on the other. One important message of the *ntoo xeeb* movement to the Hmong community itself was the need for being responsible stewards of the environment. Thus the conflict between Hmong and lowland farmers was translated into the first modern articulation of Hmong watershed management. This message was in part a response to lowland claims that the Hmong, as shifting cultivators, could not manage resources sustainably because they have no deep connection to the land. While these developments unfolded at the level of public discourse, Hmong people in other areas were working to secure access to resources at a more local level, by creating a new sense of local community and establishing new regimes of upland water management.

Managing Watersheds: Water Use and Local Institutions

The Chom Thong conflict and ensuing Hmong efforts to transform the public perception of their agricultural and resource management practices illustrate the interplay between the legal rights of resource tenure and reconstructed cultural practices. However, a somewhat different set of resource competition dynamics plays out on a daily basis in many places. The following section highlights how the Hmong assert control over water resources in an ongoing process of negotiation, both within their own community and with neighbouring communities.

Mae Chaem district in Chiang Mai province is one area where the transition from opium to vegetables was rapid. The result was a dramatic reorganization of the local economy and significant transformation of the landscape. The spread of cabbage cultivation in the uplands has been linked to changes in the water regime, and tensions between upland and lowland communities have been acute. The experience of Ban Phui Nua, a village located in the uplands of Mae Chaem district, shows how the Hmong used the cultural symbols and ritual practices mentioned above, to assert control over resources and manage the socio-economic changes going on around them.[4]

The Hmong first settled in Ban Phui Nua in 1974. The ancestors of this group left southern China around the 1890s. History blends with legend three generations after the founders of the village, but their account of leaving a place called Paaj Tawg Laag[5] because of heavy tax burden and land shortage fits with the general history of Hmong migration into Southeast Asia. The Ban Phui Hmong entered Thailand somewhere in the mountains west of Mae Sai, passing through Fang district and gradually making their way into Mae Chaem district. Villages were moved frequently, in search of land suitable for opium poppy, which was a key component of their agricultural system. When forests were abundant, old growth forest was cleared for several years' cultivation, at which point the village would move to a new site. When Ban Phui was established, government efforts to eradicate opium production, conserve upland forests and contain communist activities were placing pressure on the Hmong to settle permanently and adopt cash crop alternatives to poppy (Renard, 2001).

The establishment of Ban Phui Nua as a permanent settlement brought the villagers into a new era of village governance (Badenoch, 2008). Cooper (1979) asserted that the concept of a village as a permanent settlement did not exist in Hmong society, but rapid integration into the Thai administrative state and economy meant that the basic assumptions of Hmong village life would be challenged. While in the past moving was always an option for Hmong villagers, the newly established villages faced a new urgency to manage their internal affairs in one geographic setting.

In the post-opium era competition for land brought new tension to the community as the village grew. The rapid expansion of dry season vegetable cropping starting in the late 1990s resulted in conflicts over land and water. The new tensions were experienced internally among farmers and with neighbours in their watershed as well. In response to these changes, the Ban Phui Nua Hmong developed a new set of institutions to deal with these local issues. These innovations mirrored some of the developments made with the Hmong Environmental Network introduced above, but were more inwardly focused.

Creating watershed forests

When they arrived at the current village site in 1974, the leaders of the Yaaj clan were already aware that they had reached the final destination in their journey from China. After the move to Ban Phui Nua, village leaders began to devise a strategy to enable permanent livelihoods in the new site. The area was previously used for poppy cultivation, but in the mid 1980s the village began a transition to vegetable cultivation in upland fields. Under pressure to stop poppy production, the farmers experimented unsuccessfully with a range of alternative crops, recommended by the government and international development agencies, until they found a market opportunity in cabbage. Cabbage production took off as the main economic crop establishing permanent cultivation of sloping lands that required high chemical inputs to compensate for losses in soil fertility.

The Hmong have become well-known for this agricultural transformation and the impact it has had on the upland landscapes. However, the first large communal undertaking in Ban Phui was of forest restoration. The area above the village was dominated by *Imperata* grass (*nqeeb*). According to Hmong beliefs, the land above the village should be forested. Tree cover gives the village a 'cool' environment and helps ensure a reliable supply of clean water. The Hmong also believe that a dry and hot village is prone to disease. Villagers describe in detail the process of change in which the community protected forest area was rejuvenated:

- at first the area above the village was dominated by *nqeeb* grass, making it useless for agriculture;
- village leadership made the decision to protect the area, and devised a simple system of use regulations including the establishment of the *ntoo xeeb* as a single tree in the small area of remaining forest;
- the villagers built and maintained firebreaks, protecting against fires to allow small vegetation to grow;
- in the early 1980s, a weed called *nroj communist*[6] spread across the landscape. This weed smothered the *nqeeb*, and the lack of fire events allowed small trees and other vegetation to mature;
- once the trees reached a viable size, livestock were grazed in the area, providing natural fertilizer to the soil;

- by 1982, the vegetation had reached a size such that the *ntoo xeeb* 'area of influence' could be broadened to be a watershed forest for the village;
- more formalized regulations were created to control use of land and resources around the *ntoo xeeb*.

Throughout this time, the area of old forest in the village decreased as more land was brought under cultivation. However, the *nqeeb* grassland and the areas of older fallow were returned to forest. Although there was indeed a net loss in forest area, the landscape was reorganized by the Ban Phui Nua villagers. The forest area was effectively relocated over 20 years to the critical watershed area above the village.

Ntoo xeeb was a key component to ensuring the success of the forest rehabilitation, because the practice of *ntoo xeeb* recognizes the *thwv tim*'s influence in the forest surrounding the tree, creating a sacred space. According to traditional practice, cutting trees in the general area of the *ntoo xeeb* was prohibited. In the early days of Ban Phui Nua settlement, however, the *ntoo xeeb* was not associated with a clearly defined, wide expanse of forest. This was partially for the practical reason that there was not much remaining forest in the area. Village leaders used the *ntoo xeeb* as a tool to provide legitimacy and a cultural foundation to control fire and regulate forest use. At this point in the village development, farmers were eager to scale-up their cabbage production, meaning that access to land and fallow forest was critical and encroachment on the village forest was a real concern.

The RFD implemented a reforestation programme in the adjacent area. The programme planted pine trees in the name of watershed conservation, although the Hmong speculate that the pine trees consume more water than they provide. The planted forest is now classified as subsistence use forest, but villagers can make very little use of that forested area. The RFD forested area is now known by its Thai name as *pa anurak chumchon* (community protected forest), but the village watershed forest is called by the Hmong term *hauv dlej* (water source). This community-defined watershed area was the source of the domestic water supply (and would become a source of irrigation water in the future) and was soon recognized by the RFD as a protected forest.

The main water source for Ban Phui Nua was in the rehabilitated forest area. Mountain streams were piped to the village to provide for daily water needs. Protection of the forest area around the main water source (*qhov dlej txhawv*) is common in Hmong traditional practice, and has been promoted by the Hmong Environment Network in conjunction with *ntoo xeeb* rituals in other areas (Wanitpradit, 2005). In Ban Phui Nua, the traditional practice of propitiating the watershed spirits (*xyeem hauv dlej*) at the village source has all but disappeared and village elders say that in the past the ritual was conducted as needed, usually when there was water shortage from lack of rains. But now *ntoo xeeb* has come to be the main expression of the village's dependence on nature.

Ntoo xeeb and village cohesion

'May we all live together peacefully and prosperously, without problems or disputes, without disease or hardship'; Ntsuab Pov Yaaj, the ritual performer responsible for the annual *ntoo xeeb* ceremony, had just made the required offering of two roosters, incense, decorative paper and maize whiskey that secured the relationship between the community's members and the local spirit (*thwv tim*) for another year. As discussed above, the *ntoo xeeb* has become a symbol of environmental awareness in the Hmong society. However, the 2004 ceremony ended with a call for harmony within the village, and particularly between the three Hmong clans that live in the village.[7] This was followed by the leaders of the Yaaj, Nkws and Tsaab clans, who paid their respects to Ntsuab Pov and led the younger participants in paying their respects to the elders of all three clans. The younger men then prostrated themselves in front of the local spirit and the elders, offering maize whiskey and their commitment to work together throughout the next year (Box 4.2).

Underlying the messages expressed in this ritual was growing tension over natural resource use. By 2000, the disputes over water use and watershed management, both within the village and with neighbouring villages, had exacerbated the pre-existing tensions among the clans. The *ntoo xeeb*, the strongest symbol of village solidarity based in the Hmong cultural heritage, became increasingly important. For the traditional village leadership, it was important that the *ntoo xeeb* continue to provide a common point of reference to hold the village together socially.

In practice however, even this shared cultural symbol was not immune to social differentiation within the community. In the weeks preceding the ceremony, several Tsaab and Nkws farmers had mentioned that they might not attend the ceremony because they felt it belonged to the Yaaj, who are politically dominant in the village and have been the clan that perform the *ntoo xeeb* ritual since the founding of the village. Yet, village elders were unanimous in saying that all male members of the community should participate in the ceremony, and this sentiment was shared by the vast majority of the participants. However, during the walk to the ceremony site, one young Yaaj man remarked how few Tsaab were present in the group. 'Don't they care if the price of cabbage goes down again this year?' he wondered, implying the direct influence the local spirit is seen to wield over village affairs.

Ban Phui Nua internal developments continue to be important for decisions over natural resources. Village governance had become increasingly complex, with different spheres of decision-making authority overlapping in village life. First of all each clan had its own traditional leadership (*tug coj noj coj ua*), which was the main point of reference for decision-making beyond the household level. In addition to this 'informal' leadership, the official office of the village headman held authority over matters concerning the state bureaucracy, including relations with neighbouring villages. In recent years, with a national decentralization policy that aimed to strengthen democracy at the sub-district level in the early 2000s, elected local officers had begun to assume a role in the village, as they were involved in

Box 4.2 The *ntoo xeeb* ceremony in Ban Phui Nua

The *ntoo xeeb* is where the villagers pay respect and make offerings to the *thwv tim*, the supreme spirit of the land. Yaaj clan elders tell that they have had *ntoo xeeb* in their villages in the past, but the practice was not institutionalized. If there was illness or other social problems, the ritual leader would choose a tree and make an offering (*teev*) to the *thwv tim*. The Tsaab and Nkws elders say that they have never had a specific *ntoo xeeb* but have made offerings to the *thwv tim* as needed. When the three clans established Ban Phui Nua, the elders, led by the ritual performer Ntxoov Teev Yaaj, agreed that there should be a fixed place for villagers to make annual offerings to the *thwv tim*. The first site of the *teev thwv tim* was at a cluster of large boulders a 30-minute walk up a steep slop on the mountain behind the village. This information is interesting as Leepreecha (2004) notes Vietnamese Hmong elders remember that in the past Hmong in Vietnam used to perform sacrifices to the *pob zeb xeeb*, spirits in large rocks placed around the village, in addition to the *ntoo xeeb*. Tran (1995) confirms this, explaining that the *teev thwv tim* can be performed at a large tree or large rock. The ceremony is an important symbol of village solidarity in Vietnam but is not widely practiced in Thailand.

After eight years, incidence of illness in the village increased suddenly. It was decided that the site for *teev thwv tim* was too hot, causing disease to spread among the villagers. The site was then moved down the hill closer to the village. This time, a tree was chosen to be the first *ntoo xeeb*. It was not uncommon for a Hmong village to institutionalize the *ntoo xeeb* in response to a major village problem (Leepreecha, 2004). The situation improved, but when a large branch broke off of the tree it was regarded as a bad omen. Ntxoov Teev believed that this was because the *ntoo xeeb* was too close to the village, so it was moved a distance back up the hill towards the rejuvenating forest. The ceremony was performed at this site until 2002, when villagers began to worry that a project to construct a football field in the immediate area of the *ntoo xeeb* would upset the *thwv tim*, the site was shifted finally to the current spot, further up the slope towards the original site. In 2004, the walk up the hill became too difficult for Ntxoov Teev, and the duties for performing the ceremony were passed to Ntsuab Pov. Just after the 2004 Asian tsunami disaster, villagers noticed that a large crack had opened in the former *thwv tim* boulder, which was threatening to dislodge and roll down the mountain onto the Tsaab-Nkws settlement site. There was some talk of the possibility that the *thwv tim* was upset by some village happening, but the most immediate concern was for securing assistance from the District to build a support structure to protect the village.

allocation of sub-district rural development funds.[8] In one land conflict with a neighbouring Karen village, representatives of all three of these spheres of authority were present at the negotiations.

Ban Phui Nua falls within the Southern Network of the Hmong Environmental Networking Group, but the clan elders had already seen the need to establish and give new meaning to the *ntoo xeeb* tradition in order to meet immediate challenges of creating a sense of community within the village and bringing an awareness of ecological linkages to members of that community.[9] One area of contention among

Ban Phui Nua villagers since the beginning of opium poppy replacement was the degree to which conservation should be a part of the village landscape. Opinions were articulated in terms of clan strategies, and there was a significant degree of discord with regards to how the village should deal with internal growing pains as households started to compete for land. Dealings with neighbouring villages brought added tension to the internal dynamic, as friction with Karen neighbours over land, forest and water intensified.

The leadership of Ban Phui Nua was able to use the *ntoo xeeb* for two purposes: enhancing the environmental conditions upon which the economic development of the village would rely, and strengthening the sense of community cohesion in the village. But tensions persist within both of these projects: the debate about forest and land management continues as farmers desire more land for agriculture, and there are some doubts about the strength of the *ntoo xeeb* ceremony as a unifying force in the village. Thus the *ntoo xeeb*, as a village cultural institution, is steeped in seeming contradictions that stem from tensions within the Hmong clan system, where cohesion within clans can be accompanied by competition between clans (Badenoch, 2006, 2008). In this case, it could be argued that the village leaders were able to use the *ntoo xeeb* to counter pressures associated with the first steps of moving out of poppy production, unifying somewhat fractious groups in the face of an external threat.

In this way the forest above Ban Phui Nua was actively rejuvenated by the village customary leadership, based on a dual approach combining traditional practices and local environmental knowledge. The motivation was a combination of concern for the village water source and growing pressure from the government and lowland society. The key transformation in the process was moving from *ntoo xeeb* as a single tree, to *ntoo xeeb* as the centre of a larger patch of forest imbued with spiritual meanings and authority for regulation of social affairs.

Sprinkler irrigation and institutions across ethnicity

The largest shock to the upland natural and social environments was the introduction of sprinkler irrigation for dry season vegetable production. These gravity-fed sprinkler systems rely on small mountain streams. Small wooden and rock dams impound water, which is transported by polyvinyl chloride (PVC) pipes to upland fields, often by way of mid-point collection tanks and ponds, with no energy inputs beyond the farmers' labour. The engineering practiced by local farmers can bring water from as far as 10 or more kilometres, as the pressure created by steep slopes at the headwaters enables farmers to lift water and accommodate natural barriers in the landscape. Pipes are frequently buried in the earth, which gives a certain degree of security, but can make monitoring leakages difficult.

The first years of sprinkler irrigation (mid 1980s) in Ban Phui Nua were characterized by individual investment in infrastructure in an upland stream, managed at the household level or among close relatives within a single lineage group. But these systems were soon to become increasingly complex, with numerous extraction pipes in the source streams and multi-user storage facilities in the fields. The increasing competition for water, as the number of users in the system grew, meant that more sophisticated management arrangements were needed. Water-sharing arrangements, water use rotations, joint infrastructure investments and problem-solving committees have developed to meet the needs of upland irrigators. But the experience has been mixed in Ban Phui. On one hand, the Hmong farmers expanded their basic unit of collaboration from the household to include multi-clan water-sharing arrangements in shared irrigation systems. On the other hand, access to water is still fraught with tension, as the problem-solving mechanisms at the village's disposal are limited in their effectiveness. Thus, technology and institutions have evolved in tandem, as upland farmers are linked in new ways through ecological interactions in small upland catchments (Badenoch, 2009).

The introduction of sprinkler irrigation has brought a new layer to the relationships among upland communities as well. Watershed conflict is most commonly conceived of as a tension between fundamental differences between uplanders and lowlanders: differences in history, language, livelihood, belief systems and traditional practice (Wittayapak, 2008). But beyond the broad discourse of state versus community, upstream versus downstream and Thai versus minority, there is a complex set of dynamics evolving among upstream users themselves, who are often culturally as different from each other as they are from the lowland Thai. In the past, competition for land and access to forests were a part of upland inter-ethnic relations. But other interactions included exchange of goods and labour, intermarriage, and participation in ritual practices. The spread of sprinklers in Ban Phui Hmong was part of a larger context, in which they increasingly came into competition with their Karen neighbours. However, ritual symbols such as *ntoo xeeb* would not be helpful in establishing a new regime of watershed management with the culturally different Karen.

The Huai Sai Khao valley,[10] located between Ban Phui Nua and San Pu Loei, is the site of a struggle to establish new institutions to manage the increasingly scarce water resources that enable farmers' production of shallots and other dry-season vegetables. The Sai Khao stream springs from the rock face in a forested area above the cultivated area. This area is treated as community watershed forest by the Karen of San Pu Loei village, and has traditionally been the source of water for a small area of irrigated paddy. The PVC irrigation pipes expanded locally starting in the late 1980s, and in the 2004–2005 planting season there were 20 withdrawal pipes in the Sai Khao stream. Of these, 14 belonged to the Hmong, while the Karen owned 6. Pipes are usually shared by two or more farmers, as there is a general agreement between both groups that strictly individual use of pipes is inappropriate. Farmers sharing abstraction pipes jointly maintain the weir and pipes. Most pipes were originally installed by one farmer, but access was subsequently granted to one or more farmers as more people began dry season cultivation. Figure 4.2 provides a schematic representation of the Huai Sai Khao irrigation system. The numbered

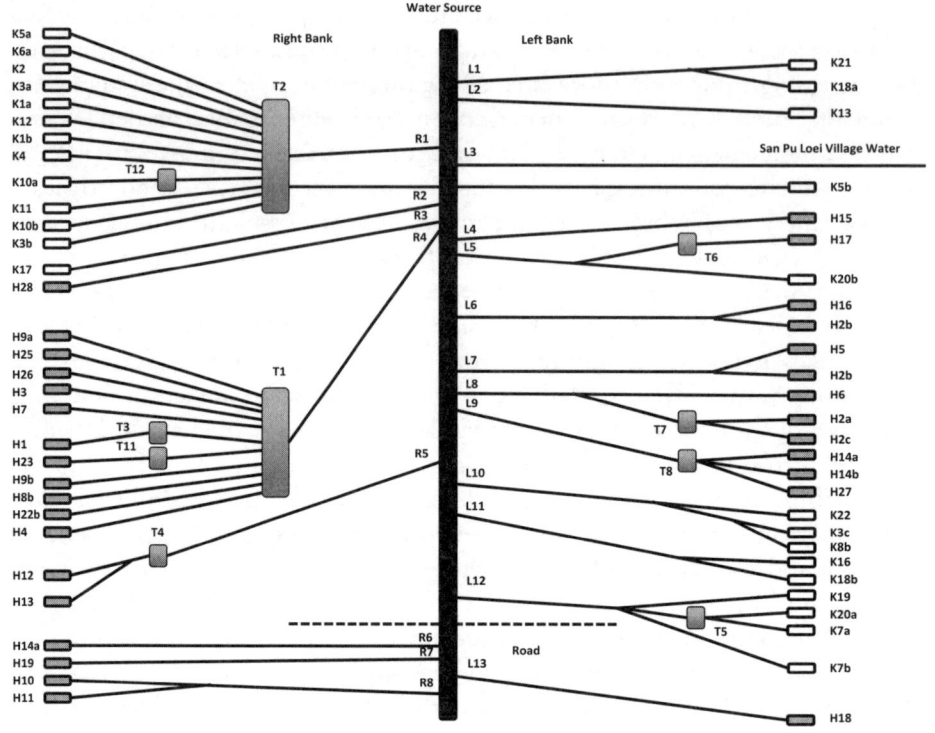

Figure 4.2 *Huai Sai Khao irrigation*

Source: Badenoch, 2009

blocks on either side are individual shallot fields, fed by pipes and intermediate storage tanks; Karen fields are white, Hmong fields are grey.

As competition for water increased, some farmers in both the Karen and Hmong communities were able to increase the scale of water management to allow more storage capacity and more sophisticated systems of water allocation. As shown in Figure 4.2, the two large water storage tanks (labeled T1 and T2) on the right bank were created to increase cooperation but did not produce collaborative arrangements between the Karen and Hmong. In the whole system there is only one example of Hmong–Karen cooperation. Nevertheless, the storage infrastructure has reduced tension among users using the shared storage tanks. However, it did not solve the problem of competition between users within the valley because there was no system-level water allocation or management mechanism. In the face of continued tensions between farmers, the traditional leaders of the Hmong and Karen community created a committee to mediate conflicts within the valley. The two Hmong and two Karen elders deliberated to solve the unregulated expansion of water extraction. In the end, despite the efforts of the committee, the local farmers were unable to implement a process of negotiated water management. The reasons offered by farmers in the valley were that there was no official support from the village for the committee's decisions, and that the lack of trust among farmers prevented a broad acceptance of the need for negotiation and dialogue.

The Karen and Hmong have been managing irrigation water with this technology for over 10 years. But because this is a relatively new addition to the livelihood technologies, there are no specific customary institutions for irrigation management in upland fields. The Karen had experience with managing small-scale irrigation for paddy production, but the Hmong were much less familiar with these practices. Nevertheless, local institutions to manage the physical and social dynamics of this new technology have emerged and are continually adapting. The institutions observed here encompass two types of evolution. First, existing social institutions are adapted to meet the needs of managing scarce water resources. Second, new institutions are created to address the increasingly complex social relations encompassing the agro-ecosystem. It is interesting to note that there is only one case of water sharing between the Karen and Hmong in Huai Sai Khao. This arrangement is actually a water-for-land swap, but was only valid for one season. When asked why there has been no direct Karen–Hmong collaboration, farmers from both sides mentioned that the lack of trust between the two groups prohibits any concrete action in joint investment or allocation negotiation.

For example, the water users of the Hmong storage tank have begun basic land-use planning to limit the number of users at the beginning of the dry season. In the 2004–2005 season, it was agreed that one user would be excluded from access to irrigation water because he had large landholdings in other areas and his claim on water in the Huai Sai Khao valley would upset the overall equity of access to the scarce resource. All farmers in the valley, both Karen and Hmong, agree in principle that such land-use planning at the stream level is necessary, although it has not yet become a reality. In other areas, one first step to controlling demand has been limiting the number of sprinklers allowed for each parcel of land. The farmers of Huai Sai Khao have not tried this mode of regulation. In many Hmong villages in the area, village water committees charge households for domestic water consumption, based either on a flat rate per household or by the number of household members. Irrigators in the Huai Sai Khao valley are managing their water at several levels, from the individual to larger user groups.

The development of irrigation technology was accompanied by institutional development, in which farmers created more complex management arrangements to deal with growing competition for scarce water. The key challenge in this institutional development was how to step outside of the customary boundaries of cooperation. At the village level, this meant moving beyond the strictly kinship-based networks of cooperation preferred by the Hmong. At the larger landscape level, the new institutional arrangements were challenged to bring together Hmong and Karen farmers in a shared regime of resource access and management (Badenoch, 2009). Indeed, there is an underlying ethnic component to this tension, as much of the conflict is described locally in terms of Karen and Hmong management practices. However, a review of the history of irrigation infrastructure development in the Huai Sai Khao valley shows that the limits of local institutions to deal with the conflict go beyond ethnic differences. Beyond the village level, the evolution of practical management arrangements in sprinkler irrigation systems is a dynamic illustration of the need for creating shared institutions that deal with conflict between communities. From the experience of Huai Sai Khao we see that some sort of facilitation may be necessary to catalyse cooperation between villages.

Discussion: Strategies for Resource Access at Multiple Scales of Governance

Local people create collective rights over forests, drawing on local values and meanings as a framework for establishing shared norms and then into more formal practices (Ganjanapan, 2000). This local articulation of rights over forests is similar to how rights over irrigation are developed as 'institutions as rules in use' (Ostrom, 1992). Conflicts over water use in upland irrigation systems, in this case between the Hmong and the Karen, have transformed the local discourse from a matter of forest management to a focus on water management allowing the actors to address the governance issues at stake directly. But tension at the larger watershed scale, pitting lowland Thai society in opposition to upland minority society, is defined by concerns over forest management despite the fact the impacts of upland land use on the hydrological regime are the damages claimed by lowlanders. In this discourse, uplanders are not recognized as water users, despite the complex technologies and management systems that have evolved. Moreover, in the research area, proposals to measure flow levels in the stream in lowland villages were rejected by the lowland villages themselves, most likely because lowland farmers themselves have drastically increased their demands for water during the dry season.

It has been suggested that the diversity of local actors and the water management systems they are creating requires a flexible legal framework (Sangkapitux et al, 2006). Pluralistic legal frameworks that can be flexibly applied in recognition of the diversity of water use institutions that have evolved in the uplands will likely be a key element in local water governance (Neef et al, 2006). Indeed, within the socio-cultural and political biases aligned against uplanders in the Thai watershed discourse (Laungaramsri, 2001), the fundamental question needs to be asked: how do upland sprinkler irrigation systems fit into the larger resource governance landscape? The short answer is that they do not have clear linkages to institutional authority at higher levels. The Royal Irrigation Department does not have jurisdiction in these areas, as they are classified by the RFD as 1A Watersheds, where the restrictions on land use are the strictest. Human activities are legally not allowed, and the RFD is responsible for enforcing land-use restrictions and forest management regulations. Even if the government has had to recognize de facto

Table 4.1 *Strategies employed by the Hmong*

Level	Rights approach	Rites approach
Village Governance	Devise practical resource access rights based on new technologies	Strengthen and adapt customary practices to increase village solidarity and mediate resource competition
Catchment Management	Devise formalized regulation of forest and land use to bridge inter-village and inter-ethnic group tensions	Construct informal arrangements for dialogue around water, with traditional leaders playing key role
Environmental Discourse	Assert traditions of environmental awareness to enhance legitimacy in public perception	Raise internal awareness of responsibilities in and capacity for resource stewardship

a certain level of human settlement in restricted watershed areas, without secure land rights it is difficult to imagine how a regime of water rights could be devised in the official legal arena.

The dynamics of diversity vary as one moves from the local to national and regional scales. Bringing analysis of multiple scales to the challenges of governance within watersheds can shine light on the institutional options that may be appropriate for a situation of diversity. However, questions of how cultural dynamics play out in a watershed may require a different lens (Lebel et al, 2008). In their theory of access, Ribot and Peluso (2003) describe a framework that moves beyond the equation of resource access with property rights, and broadens the tools of access from 'bundles of rights' to 'bundles of power'. This means looking at access – people's ability to benefit from resources – as the full range of tools that communities use to generate and assert power over resources. Opening up the analysis to a wider range of social relationships through which local actors obtain access is essential in a culturally diverse setting such as the multi-ethnic upland watershed areas of northern Thailand.

The Hmong strategies described here represent an effort to bridge the gap in linkages across levels of governance. At each level of governance, the Hmong have used rights and rites to reinforce each other. Table 4.1 summarizes the approaches taken at the levels of village governance, catchment management and environmental discourse. The village-level rights approach seeks to expand customary kinship networks to achieve a practical and predictable water management regime to govern the new irrigation technology. At the same time, the village-level rites approach represents a formalization of customary practices pertaining to the position of watershed forests in the village. Efforts to establish institutions at the catchment level have had mixed results, but a common theme is the focus on forest and land management issues rather than the water use issues that drive conflict. At the level of environmental governance, the Hmong assert traditions of environmental awareness in order to create a legitimate space for themselves in their landscapes and in the public eye. The rites approach at this level also is an internally oriented call to strengthen the Hmong's own sense of responsibility and capacity in their own livelihood activities.

Resource conflict is a particularly rich arena in which local actors broaden the range of 'policy mechanisms' that they use to increase access. Power can arise from social identity and negotiation of other social relationships, as seen in the strategies taken by upland communities in Thailand such as the Hmong. Efforts to assert power to secure access to water can be seen at several levels of governance, reflecting the nested scales of management within a watershed. At the same time, use of multiple scales of governance reflects an important strategy for creating bundles of power.

There are strengths and weaknesses to both the rights- and rites-based approaches. The diversity of the area is itself at the same time an opportunity and a constraint, in that the rich resources of ritual and cultural practice can have powerful meaning for members of the community, but may provide little in the way of creating common ground for negotiation between ethnic groups. Regulatory institutions created around infrastructure have drawn on the economic necessity of equitable water management to deal with conflict between groups, but because they remain informal there is no mutually recognized source of authority to provide the necessary facilitation or mediation. Formalized institutions must be supported by a certain degree of trust and shared understanding of the problems.

Conclusions

The redefinition of cultural heritage to meet the demands of watershed management has been an effective way for the Hmong to engage in the policy-level discourse on their own terms. In response to the legal denial of legitimacy as managers of watershed resources, the Hmong have asserted a cultural heritage reflecting their own understanding of human, forest and water relationships. This is an important step on the long road towards securing rights to manage watershed resources, and should be understood as one part of a larger local response to the forces of globalization and regionalization (Leepreecha et al, 2008).

The same cultural institutions are being redefined and recreated at the local level in an effort to deal with the daily challenges of managing watershed resources. The primary concern at this level is to establish internal capacity to deal with internal conflicts, as villages move into new areas of resource governance. For the Ban Phui Nua Hmong, the creation of 'community' is a struggle to establish a balance between cooperation and competition. In this case, the redefinition and assertion of the *ntoo xeeb* was done without explicit reference to the larger Hmong movement, but was intimately intertwined with the fast-paced economic and landscape changes that began with the end of poppy production. At both levels, there is a mixture of rights and rites-approaches.

Local cultural institutions in the study area play a more limited role within contexts of multi-ethnic landscapes. The Hmong and Karen communities have devised ways to collaborate on the daily challenges of water management through the joint creation of new institutions. These institutions are oriented towards shared understanding of 'rights' of access, rather than practice of 'rites' although the main mechanism in mediating competition between the groups is a tenuous link between customary leaders of the two communities. Although the broad application of cultural resources seems to have been of limited interest and effectiveness, the need for both practical institutions for resource management and solid networks of trust in implementing them suggests that creation of shared local politico-cultural spaces may still have potential in the ethnically diverse upland areas.

Watershed management is an interaction of ecological and human systems at multiple scales. There is a tension between culturally specific institutions and the need to establish institutions that can bridge the gaps between diverse user communities. Watershed governance in ethnically diverse areas will benefit from approaches that combine a concern for both rights-based and rites-based approaches. The strategies being developed show how local people are not limited by discourse dichotomies that divide the world into the extremes of nature destroyer or protector. Similarly, these approaches avoid simplistic legalistic and cultural heritage oppositions, while highlighting emerging local views on community and state. It is the ongoing negotiation of these diverse factors that expresses the local adaptations that enhance resilience of local people while they create and expand spaces for themselves in the upland landscape of watersheds.

Notes

1 See, for example, the International Association for the Study of Common Property Resources' Digital Library of the Commons, available at http://dlc.dlib.indiana.edu/dlc/ (last accessed 26 February 2010)
2 See, for example, Ganjanapan (2000).
3 Leepreecha (2004) and Wanitpradit (2005) provide detailed accounts of the *ntoo xeeb* ceremony and its transformation.
4 The data used in this analysis was collected between 2002 and 2005.
5 Wenshan, Yunnan Province, China.
6 *Chromalaena odorata*, a successional plant in shifting cultivation areas, appeared in the village during the years (1970s) when communists were taking to the forest in the area around Ban Phui Nua.
7 The basic unit of social organization in Hmong society is the clan, a kinship group of shared descent that provided the necessary resources for securing access to and managing the natural resources upon which their livelihoods depended. Hmong clans are differentiated by surname, ritual practices and food taboos. Hmong people must marry outside of their clan, which means that while kinship networks represent the strongest source of social cohesion, these networks are constantly being expanded by marriage.

8   The issue of decentralization is discussed in depth in C. Wittayapak and P. Vandergeest (2010) *The Politics of Decentralization: Natural Resource Management in Asia*, Mekong Press, Chiang Mai.
9   The role of the *ntoo xeeb* in alleviating Ban Phui Nua inter-clan tensions seems similar to the situation in Ban Mae Sa Mai, where several Christian sects live together with traditional Hmong ritual practitioners. Maintaining internal cohesion among followers of different ritual and spiritual traditions proved to be a challenge for the village leadership.
10 The evolution and management of the irrigation system is presented in more detail in Badenoch (2009).

References

Badenoch, N. (2006) 'Social Networks in Natural Resources Governance in a Multi-Ethnic Watershed of Northern Thailand', unpublished doctoral dissertation, Kyoto University Graduate School of Asian and African Area Studies, Kyoto

Badenoch, N. (2008) 'Managing cooperation and competition: Hmong social networks and village governance', in P. Leepreecha, D. McCaskill and K. Buagaeng (eds) *Challenging the Limits: Indigenous Peoples of the Mekong Region*, Mekong Press, Chiang Mai

Badenoch, N. (2009) 'The politics of PVC: Technology and institutions in Thailand upland water management', *Water Alternatives*, vol 2, no 2, pp269–288

Cooper, R. (1979) 'The Yao Jua relationship: Patterns of affinal alliance and residence among the Hmong of northern Thailand', *Ethnology*, vol 28, pp173–181

Cooper, R. (1984) *Resource Scarcity and the Hmong Response: Patterns of Settlement and Economy in Transition*, Singapore University Press, Singapore

Cornford, J. and Matthews, N. (2007) *Hidden Costs: The Underside of Economic Transformation in the Greater Mekong Subregion*, Oxfam Australia, Victoria

Daniel, R. and Lebel, L. (2006) 'Land policy, tenure and use: Institutional interplay at the rural-forest interface in Thailand', USER Working Paper WP-2006-1, Unit for Social and Environmental Research, Chiang Mai University, Chiang Mai

Forsyth, T. and Walker, A. (2008) *Forest Guardians, Forest Destroyers: The Politics of Environmental Knowledge in Northern Thailand*, Silkworm Books, Chiang Mai

Ganjanapan, A. (2000) *Local Control of Land and Forest: Cultural Dimensions of Resource Management in Northern Thailand*, RCSD Monograph Series No 1, Regional Center for Social Science and Sustainable Development, Chiang Mai

Hirsch, P. (1990) *Development Dilemmas in Rural Thailand*, Oxford University Press, New York

Laungaramsri, P. (2001) 'The ambiguity of "watershed": The politics of people and conservation in northern Thailand', *Sojourn*, vol 15, pp52–75

Lebel, L., Daniel, R., Badenoch, N., Garden, P. and Imamura, M. (2008) 'A multi-level perspective on conserving with communities: Experiences from upper tributary watersheds in montane mainland Southeast Asia', *International Journal of the Commons*, vol 2, no 1, pp127–154

Leepreecha, P. (2004) '*Ntoo xeeb:* Cultural redefinition for forest conservation', in N. Tapp, J. Michaud, C. Culas and G.Y. Lee (eds) *Hmong/Miao in Asia*, Silkworm Books, Chiang Mai

Leepreecha, P. (2008) 'The role of media technology in reproducing Hmong ethnic identity', in D. McCaskill, P. Leepreecha and S. He (eds) *Living in a Globalized World: Ethnic Minorities in the Greater Mekong Subregion*, Mekong Press, Chiang Mai

Leepreecha, P., McCaskill, D. and Buadaeng, K. (2008) 'Introduction', in P. Leepreecha, D. McCaskill and K. Buadaeng (eds) *Challenging the Limits: Indigenous Peoples of the Mekong Region*, Mekong Press, Chiang Mai

Na Ayuthaya, P.N. (1996) 'Community forestry and watershed networks in northern Thailand', in P. Hirsch (ed.) *Seeing Forests for the Trees: Environment and Environmentalism in Thailand*, Silkworm Books, Chiang Mai

Neef, A., Chamsai, L. and Sangkapitux, C. (2006) 'Water tenure in highland watersheds of northern Thailand: Managing legal pluralism and stakeholder complexity', in *Institutional Dynamics and Stasis: How Crises Alter the Way Common Pool Resources are Perceived, Used and Governed*, Regional Center for Social Science and Sustainable Development (RCSD), Chiang Mai University, Chiang Mai

Ostrom, E. (1992) *Crafting Institutions for Self-Governing Irrigation Systems*, ICS Press, San Francisco

Peluso, N.L. (2005) 'From common property resources to territorializations: Resource management in the twenty-first century', in P.R. Cuasay and C. Vaddhanaphuti (eds) *Commonplaces and Comparisons: Remaking Eco-political Spaces in Southeast Asia*, Regional Center for Social Science and Sustainable Development, Chiang Mai

Renard, R.D. (2001) *Opium Reduction in Thailand, 1970 to 2000: A Thirty Year Journey*, Silkworm Books, Chiang Mai

Ribot, J. and Peluso, N.L. (2003) 'A theory of access', *Rural Sociology*, vol 68, no 2, pp153–181

Sangkapitux, C., Neef, A., Nunthasen, K. and Yothapakdee, T. (2006) 'Assessing water tenure security in highlands watersheds: A case study from northern Thailand', paper presented at the Eleventh Bienniel Global Conference of the International Association for the Study of Common Property (IASCP) 'Survival of the Commons: Mounting Challenges & New Realities', 23 June 2006, Bali, Indonesia

Tran, H.S. (1995) *Van hoa Hmong*, Nha xuat ban van hoa dan toc (publisher), Hanoi

Usher, A.D. (2009) *Thai Forestry: A Critical History*, Silkworm Books, Chiang Mai

Walker, A. (2001) 'The "Karen Consensus", ethnic politics and resource-use legitimacy in northern Thailand', *Asian Ethnicity*, vol 2, no 2, pp145–162

Walker, A. (2003) 'Agricultural transformation and the politics of hydrology in northern Thailand', *Development and Change*, vol 34, no 5, pp941–964

Wanitpradit, A. (2005) '*Hauv dlej*: Watershed areas in the cultural perspective of the Hmong', research report, Chiang Mai University Social Research Institute, Chiang Mai

Wittayapak, C. (2008) 'History and geography of identifications related to resource conflicts and ethnic violence in northern Thailand', *Asia Pacific Viewpoint*, vol 49, no 1, pp111–127

5

Local Institutions and the Politics of Watershed Management in the Uplands of Northern Thailand

Rajesh Daniel and Songphonsak Ratanawilailak

INTRODUCTION

Upland watersheds in northern Thailand are arenas of social interaction and political contestation around the values, uses and management of natural resources and ecosystem services. Differences in definitions, perceptions, objectives and interests among actors abound and often lead to misunderstandings and misrepresentation of alternative land-use and watershed management practices (Laungaramsri, 1999).

Tensions in these arenas have grown in recent years with intensification of land uses and market-oriented cultivation in both upland and downstream areas. Upland farms now widely use fertilizers and pesticides and in some locations even have overhead sprinklers and associated infrastructure for irrigation water storage, delivery, and distribution (Badenoch and Wanitpradit, 2006).

The nature of the governance challenges are not dissimilar to many upland forest areas in the countries of the Mekong region as national parks and watershed conservation areas expand while upland farmers try to maintain their livelihood security and resource use.

This chapter explores how upland farmers in northern Thailand, predominantly ethnic communities, are coping with the constraints posed not only by resource scarcity such as seasonal water shortages but also by state conservation laws and official development strategies.

The local politics of watershed management in northern Thailand is affected by wider watershed management discourses, government policies related to the uplands, in particular regulations and classification systems for land, and the feedback from actual livelihood and conservation practices (see Figure 5.1).

Based on the authors' research in the Upper Mae Hae and Khun Kan watershed areas, the chapter shows that upland farmers attempt to maintain farming livelihoods by adopting and using a range of local institutions such as traditional definitions and practices as well as local government agencies. Individual actors are seen taking on greater responsibilities and roles with respect to local institutions for watershed management.

Upland farmers attempt to frame their own definitions of 'watershed' based on their cultural or customary values, transform earlier village-level institutions (Wanitpradit, 2008), drive collaboration among individual actors in positions of power, or redefine other new institutional set-ups from activities of several actors to retain control over livelihoods and upland landscapes (Lebel et al, 2008).

The chapter uses case studies in the Upper Mae Hae and Khun Kan watersheds in northern Thailand to illustrate the negotiation and contestation between the different definitions and meanings of watershed and those who use it. The cases show how the on-ground efforts of local-level actors including individual leadership, watershed networks that cut across administrative boundaries and the local administrative organization play their roles, shape their definitions and perspectives as well as further their diverse objectives in watershed management.

We think that the policy challenge for watershed governance in Thailand and the Mekong region is to provide an enabling policy framework that can be inclusive of upland – especially ethnic community – livelihoods, address resource access and scarcity and resolve resource conflicts and tensions. This chapter is an attempt to further our understanding of upland watershed management and the roles of local institutions in order to widen the options for watershed governance policies in Thailand and the Mekong region.

The chapter is structured with the following sections: outline of the key contestations over upland watersheds in Thailand including how watershed as a concept is constructed; the different ways of viewing the watershed; case studies illustrating the local institutions and watershed management practices; and discussions and conclusions.

Contested Watersheds

The backdrop for this chapter's explorations is the contemporary politics of watershed management where alternative visions for the governance of upper tributary watersheds are keenly contested. In its explorations on local politics and institutions, the chapter attempts to build on previous analysis about upland watersheds in northern Thailand, in particular by Laungaramsri (1999), showing that many tensions revolve around the 'ambiguous' ways the state has defined and represented upper tributary watersheds.

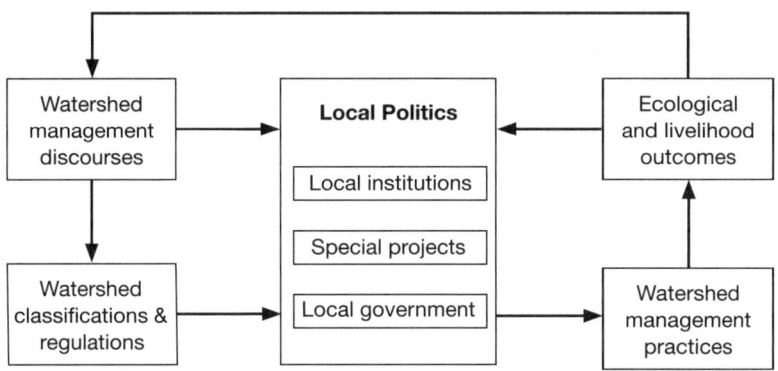

Figure 5.1 *Local politics of watershed management*

Watershed classification has often been at the core of resource conflict issues at the rural–forest interface in the uplands of northern Thailand (Daniel and Lebel, 2006). State watershed classification, centrally designed and planned with a singular vision to conserve upland forests and restrict community land uses, cannot recognize or deal with the complexities of upper tributary land-use, farming and livelihood activities of the mostly ethnic communities living and farming in the uplands (Laungaramsri, 1999).

Upper tributary watersheds also provide a range of goods and services not just to the people who live there but also others downstream as well as to society at broader spatial and multiple temporal scales (Lebel et al, 2008). Challenges to watershed management come from multi-levels and scales as upland people face pressures and opportunities as well as resource conflicts from downstream, upstream and within the community (Forsyth and Walker, 2008).

The heterogeneity of upland livelihood interests concerning resource use and benefits also affects negotiations on rules and management (Thomas et al, 2004). As roads bring villages closer to markets, most families engage in a mix of livelihoods ranging from swidden farming and gathering crops from secondary forests to planting cash crops, tourism services and migrant labour in towns and cities. Men and women, older and younger people in the same household have different views in the use and management of upland resources (Lebel et al, 2008).

Since the early 1970s, the northern Thai uplands and their ethnic communities have been the focus of attention of the Thai state and foreign donor agencies that emphasized the eradication of poppy cultivation and the integration[1] of the upland ethnic communities. These efforts promoted 'highland development' through expansion of cash crops and the building of roads and schools (Puginier, 2002; Thomas, 2002). By the 1980s, around 168 agencies from 31 government departments and 49 international donors were involved in highland development (Ganjanapan, 1997). Meanwhile, increasing concerns[2] over the loss of upper

Table 5.1 *Areas of different land according to the watershed classification scheme of Ministry of Science and Technology in 1996 (adopted by the RFD)*

Class	Definition	Land-use	Total area (km^2)	% All
Class 1A: Permanent Forest Cover	Very high elevation and very steep slopes	Protected or conservation forest and headwater source	85,464	16.7
Class 1B: Permanent Forest with already cleared areas	Very high elevation and very steep slopes	Should be reforested or maintain in permanent agroforestry	7627	1.5
Class 2	High elevation and steep up to very steep slopes	Commercial forest	42,769	8.3
Class 3	Uplands with steep slopes	Fruit tree plantation	39,284	7.7
Class 4	Gentle slope areas	Upland farming	81,284	15.8
Class 5	Gentle slopes, flat areas	Lowland farming	251,484	49.0
Reservoir area			5455	1.1
Total			531,367	100.0

Source: Chankaew and Team Ae (1996); Knie and Moller (1999)

tributary forests led to Thailand's Royal Forestry Department (RFD) formulating a 'watershed classification' system in 1983 of various classes of watershed viz. 1A, 1B etc. (see Table 5.1) (WCC, 1983).[3]

Classifying a land area as a watershed affords a policy tool for the state that cuts across and overrides other categories and practices of land and forest use, to justify its control of upland resources and give primacy to the state's decision about land use (Laungaramsri, 1999). The watershed classification is easy to selectively apply. It overlooks real topographic complexity when the scheme is applied in mapping exercises. Maps classify flatter areas within mountain regions as 'steep' (Forsyth, 1996; 1998). Slopes of 35 per cent or more were declared as protected watershed (class 1A). Most of the highlands were placed as watershed class 1A that prohibited any form of settlement or agricultural activity and made upland ethnic communities' livelihoods illegal (Tangtham, 1992; Puginier, 2002).

Thailand's earlier attempt to rapidly centralize forest control was the National Forest Reserve Act of 1964 that designated forestlands as state national forest reserve. By 1985, the RFD declared approximately 45 per cent of the country's total area as forest reserves. But areas classified as 'forest land' often have no trees and are already being farmed by the poor (Daniel and Lebel, 2006). This resulted in the transfer of rights from many communities who had traditionally claimed tenure in the 'forest' lands (Hirsch and Lohmann, 1989; Daniel and Lebel, 2006) and caused conflict (Flaherty and Jengjalern, 1995).

The RFD replicated these mistakes with its watershed classification in the uplands. Entire districts or sometimes provinces in the north were drawn to fall under strict conservation status. Moreover, the watershed classification can also work in tandem with the expansion of national parks and wildlife sanctuaries. In many communities, the national parks and watershed areas may even overlap bringing restrictions from both systems of control on the use of these land areas (Vandergeest and Peluso, 1995).

Although the proportions of land classified as watershed in classes with severe restrictions seem to be moderate at the national level, this proportion rapidly increases in the upland areas. For example, although only 26 per cent of the nation's land falls into Class 1 and Class 2 (the highest order of protection and thus the most limiting land use restriction categories), the proportion in these classes is twice that for the northern region and the Ping River watershed and climbs to about 90 per cent in the Mae Chaem watershed, a major tributary of the Ping River (Suraswadi et al, 2005; Thomas, 2005; Daniel and Lebel, 2006).

Most of Mae Hong Son province falls under Watershed 1A, prohibiting all settlement and farming resulting in huge stress on the communities who live and farm in the province.

The 1985 classification based on overlays of available maps for soils, topography and forest cover was used to classify each square kilometre in the watershed 'zone' into six categories (Pratong and Thomas, 1990). In practice rather modest information was available then (and even now) on soils for the 'topographically complex' areas where upper tributary watersheds lie.

Watershed: Discourse and policy

The state watershed discourse has posed restrictions on the access to these areas by upland communities and their livelihood practices; it is useful to clarify its history, motivation and consequences for socio-ecological issues and local livelihoods. While acknowledging that markets and the private sector are also important in upland watersheds, this section looks particularly at how state policies in commercial agriculture, forest land zoning, watershed classification, narcotics and national security and halting 'shifting cultivation' have resulted in changing land use in northern Thailand to a mixture of grasslands, pine reforestation, intensive commercial vegetables, temperate/sub-tropical fruit trees, etc.

At the same time, growing environmental awareness combined with increasing demands for water by agriculture, cities and industry are focusing attention on ethnic communities and their land uses in upper watersheds in northern Thailand (Forsyth and Walker, 2008). The Thai state definition of upland 'watershed' equals an area of forest that is perceived to be located at the 'headwaters' or the source of the water and people are not allowed to live there. The term watershed as used by state agencies is closely related to control and resistance. An example of policy affecting watershed management is Thailand's seventh National Economic and Social Development Plan (1992–1996) that increased the proportion of the conservation forests targeted for 1996 from 15 to 25 per cent of the national territory and reduced the area of economic forests to 15 per cent. The National Security Council, together with the Office of the Narcotics Control Board and the Royal Forestry Department (RFD) formulated the Master Plan for the Development of the Highland Community, Environment and Narcotics Control (1992–1996). The efforts culminated in providing a higher budget and a better coordination framework for relocating villagers from protected forest areas while also trying to solve drug abuse problems (Ganjanapan, 1998).

SEEING THE WATERSHED DIFFERENTLY

Authorities and local farmers see and use the landscapes very differently (Daniel and Lebel, 2006). Overall, local terminology of seeing and describing upland areas, diverse as it is across cultures, appears to pay more attention than official definitions to integrating human use of land and forests with ecological and hydrological considerations for management of upper tributary watersheds. Local watershed management practices are based on a mixture of customary beliefs, animist practice and rituals and observation and not on run-off rates, soil density and infiltration. Local meanings appear more dynamic: as illustrated in the local definitions below, there is not only the *hydrological* but also the *social* relationship (see also Chapter 4, this volume). Protection of the watershed in local terminology can also mean to participate and retain local control over managing the watershed area. The local or ethnic watershed terminology includes elements of spirituality and animism such as *phi nam* (or 'water spirit'; Wanitpradit, 2005). The local or ethnic language terms for watershed don't directly correspond with the Thai language official term of *lum nam* ('watershed'), a definition that is based on a geographical identification of an area or zone, and the term often conventionally used by state officials and foresters.

A number of examples of local classification and definition of watershed and upland forest continue to be adapted and used. Ethnic Hmong communities in the Mae Hae watershed of Chiang Mai province view a watershed forest as existing because of, and for, the village community, and involving all the people who live there and depend upon it. The Hmong define particular areas of watershed with different terms and their meanings demonstrating the inherent differences of perception of not only what a watershed is but also what it can be used for (see Chapter 4, this volume).

Similarly the ethnic Karen communities of the Upper Mae Hae watershed define the upland forest area under different zones depending on its functions, viewing the landscape as a forest-agroforestry mosaic (see Box 5.1).

> ## Box 5.1 Watershed classification of the Karen communities in the Upper Mae Hae watershed
>
> 1 *Doota*: this is a strictly no-farming area that is usually located in the upstream areas. Some of the area may also comprise areas prohibited for human use by traditional taboo beliefs. This zone falls under close community control by the Karen communities to ensure neither Karen nor outsiders can farm or use the area
> 2 *Doola*: this is the farming areas that comprise a mosaic of paddy fields, rotational farming areas and upland fallows that were formerly rice farming areas but are now recovering second-growth forest. Some areas also include water sources
> 3 *Gnaaw Ker Thor*: the area where the umbilical cord of the newly born babies are traditionally buried or ceremonially tied to trees. It's full of large trees and is considered as forest for use by the community for building houses, taking fuel wood, and holding spirit ceremonies, as well as functioning as a cemetery
> 4 *Nyi*: these comprise both the village settlement and the household garden areas.
>
> The Karen also have their own ethnobotanical zoning of upland forest ecosystems explained below with the Karen terms along with the corresponding conventional forestry classification:
>
> 1 *Ker Ner Mue* (wet evergreen forest): located at higher elevations, this is evergreen forest with a high density of large trees. The area has mild temperature, is moist all year around and does not shed leaves in the winter. A number of upland springs emerge from these areas
> 2 *Ker Ner Pa* (mixed deciduous forest): comprises both large trees as well as drier pine growth areas
> 3 *Kor Bei* (dry dipterocarp forest): with smaller (non-timber) trees that shed their leaves around the beginning of the dry season and then grow back again
> 4 *Pae Kho* (rocky soil forest): found at lower elevations having smaller trees that are very dry during the summer season, with higher temperatures, and not useful for planting as the areas are usually rocky with stony soils
> 5 *Ta Per Pue Kho* (lowland moist tropical forest): it has mixture of large and small trees, swamps and higher temperatures. The soil is usually hard and compacted in the dry season but turns red and muddy after the rains.

Local Institutions and Watershed Management in the Upper Mae Hae Watershed

A range of actors and institutions at various levels ranging from traditional leaders and headmen to water committees, watershed networks and local government bodies are involved in upland watershed management. The following sections provide illustrative case studies and profiles from two upper watershed areas of various local institutions and actors using different ways, tools and perspectives to shape their objectives in watershed management.

The Upper Mae Hae watershed in Chiang Mai province covers three districts: Mae Chaem, Samoeng and Mae Wang. The study area in Mae Hae watershed covers 15 villages comprising mainly ethnic Karen and Hmong communities in three sub-districts: Mae Na Jorn, Bor Kaew and Mae Win.

The cropping systems in the area was formerly rotational farming with poppy planting for opium; since the 1960s with state-led poppy crop replacement projects, cash cropping and fruit orchards were initiated. Since the 1980s, 'highland development projects' primarily to replace poppy crop cultivation has effected various changes to the Upper Mae Hae watershed and its land-use and farming systems with an increase in cash crop cultivation using sprinklers, PVC pipes and small ponds for irrigation. These farming changes have led to increased water competition both for domestic consumption and for commercial agriculture.

In the Upper Mae Hae, water use for domestic household consumption and irrigation has increased rapidly in the last 20 years due to the farming of cash crops and the wider use of irrigation water supply and distribution technologies. The resultant water scarcity and conflicts over resource access reach their peak during the dry season months from January to May every year. The dry season water scarcity and resulting tensions and conflicts over access to irrigation water are no longer confined within or between two individual villages but can often affect entire sub-watershed areas and many villages in both upstream and downstream locations, pitting villages and even members within households against one another.

For the Karen[4] and Hmong communities in Upper Mae Hae, watershed management is as much about protecting the forests or reforestation activities as it is about water for farming to ensure equitable and secure water use and sharing, resolving water conflicts and ensuring crops get enough water. Although external agencies such as the field unit of the RFD have pushed for land and forest conservation, villagers try in their own ways to look after individual and village-level interests over farmlands and maintaining access to irrigation water.

Traditional leaders and headmen in the Upper Mae Hae watershed

The entire Mae Hae watershed area falls under watershed class 1A and villages face the formal land-use restrictions of the forestry department. However, cash cropping has continued relatively untroubled in these villages due to the operations of various 'highland development' projects that are excluded from restrictions (see Box 5.2). For example, the Royal Highland Development Project known as *kronkaan luang* (or the Royal Project) operates in the Upper Mae Hae area. Initiated in the 1960s with the aim of replacing poppy cultivation in the highlands and supporting cash crop farming, the project farming areas receive special status from state forestry restrictions.

A significant governance challenge for the Upper Mae Hae watershed is the institutional and cultural mix as the watershed cuts across the administrative lines and hierarchies of three administrative districts as well as comprising ethnic Karen

> ## Box 5.2 Special Projects in the Uplands
>
> Special projects are development activities undertaken in the watersheds of northern Thailand. For the past 30 years, special projects have provided financial and technical resources for government agencies and upland ethnic communities. One of the most well-known is the upland development project initiated by His Majesty the King of Thailand and referred to as a Royal Project (RP). RPs were initiated in 1969 with the aim of assisting the development of ethnic communities in the upland mountains. The upland ethnic communities were (and sometimes still continue to be) viewed with apprehension by mainstream Thai society as they were at that time planting poppy (Renard, 2001) and also posed security concerns by living near or traveling freely across national borders.[5]
>
> The special projects attempted to halt opium poppy production by introducing other cash crops as replacements for poppy planting, usually temperate fruit trees or exotic vegetables. Agricultural research stations were also established to field test appropriate substitute crops. At present about 36 stations work throughout the upper north of Thailand.
>
> From 1996 onwards, several state line agencies including the RFD were integrated into the RP Working Committee, for implementing strategic plans. One of these was a land-use zoning pilot programme launched in 1998, including activities for forest management. This programme was spearheaded by the RFD through its watershed management units in the areas, in cooperation with other line agencies and the target communities. In the Mae Khan watershed for example, at present there are four RP stations that collaborate with the RFD field units and are located in Ban Khun Wang, Ban Mae Sa Pog, Ban Mae Hae and Ban Thung Luang. Usually their support for land-forest activities is available only for the communities living or farming within the RP boundary. In the Upper Mae Hae, the RP was the first upland development project to be established in the watershed. The RP entered the Upper Mae Hae area in 1986 and began a development programme in Mae Hae village that promoted cash crop farming to replace opium poppy cultivation covering the 15 villages of the Upper Mae Hae watershed.

and Hmong communities. For watershed management to be effective, actors and institutions – both informal and formal – have to work across overlapping levels. A number of actors and institutions work at different levels with particular roles, authority and duties at that level but also in relation to other levels.

Hau zos

Hau zos is the traditional leader (always male) of the Hmong ethnic community. Every Hmong community has a *hau zos* from each of the family clans. A village with more than one clan may have more than one *hau zos* but the *hau zos* carrying the most authority would come from the largest clan. The *hau zos* uses his knowledge of the Hmong ethnic ceremonies as well as his experience and skills to resolve issues and conflicts within the community and clans.

The *hau zos* has a specific role in the animist rituals in the village relating to water, forest and farming. He conducts the *fiv yeej* ritual for instance to make an agreement with the spirits of the place before cultivation or other activity such as digging ponds. The farmer requests assistance or cooperation from the spirits and then is obligated to make a sacrifice in return.

With the frequency of water-related conflicts rising, the *hau zos* has taken up a larger role in the village-level water use, working together with other *hau zos* elders as well as the village headman. He joins village meetings, offers advice and helps resolve conflicts, resorting more to cultural and customary norms than formal laws. In sum, although still acting within his bounds as a traditional leader, the *hau zos* has made himself more available to deal with and address water- and watershed-related conflicts.

Hee kho

Hee kho is the ethnic Karen leader analogous to the Hmong's *hau zos*. Each Karen community or village has a *hee kho* who is the leader of the village rituals and belief systems and known to communicate with spirits. The *hee kho* position is life long and is always an ethnic Karen male with the position being passed down from father to son or the closest male next of kin. Prior to the advent of the village headman, the *hee kho* had authority over village affairs. All Karen villages usually have one person as a *hee kho*. However, some communities such as Pha Kia Noi village in Upper Mae Hae that have recently converted to Christianity have done away with this system.

The *hee kho* has knowledge of animist and spiritual ceremonies, conducts ethnic rituals especially related to farming and uses Karen parables and teachings to resolve conflicts within the village. For instance, the *hee kho* conducts the *Lue Fai* or the water spirit ceremony for irrigation water use. Every household that takes water from the *muang faai* irrigation weir for paddy farming must do this ritual after the first *Lue Fai* of the farming season has been conducted by the *hee kho*.

The *hee kho* is also the person who keeps the oral history of the village or land and water use of the community; he is thus most often consulted for resolving or negotiating competing claims over streams or fields.

Hee kho Wattana from Huay Khamin Nok village says that he tells people to avoid conflicts over resources, 'because the forest, land and water have spirit-owners. When there is conflict or disrespect for village rules, it affects the entire village. If we have conflicts, the forests and water will disappear.'

One *hee kho* stated that although he has no direct authority over use of water, 'what the *hee kho* does is to teach villagers about how to use and share water and resources. We say, don't argue but respect the water and land spirits and the rights of each other.' While the *hee kho* primarily works on the village-level watershed issues, it is not unusual for him to call for and join in village meetings, talk to government officials or agencies about community watershed issues and work closely with the village headman as well as with external agencies including the Tambon (sub-district) Administration Organization (TAO). Thus similar to the Hmong traditional leadership, the Karen community's *hee kho* has over the recent years taken up a more proactive role in watershed management issues.

Village headman

The Thai state began formally appointing village headmen around 40–50 years ago, giving them primary responsibility for security and liaison with the central bureaucracy. The headman is elected by the population of the village and then confirmed by the Ministry of the Interior. Some of the traditional leaders' powers and responsibilities have gone to the headman while rituals usually remain under the Karen *hee kho* or Hmong *hau zos*. Village headmen perform a number of watershed management tasks including providing information to villagers about fire risks in the dry season, and matters related to watersheds and forests from regular monthly meetings at the district office. They also hold meetings in the village and are often the main conduit for official information and campaigns. They hold responsibility for coordinating and developing water supplies for drinking and domestic consumption in cooperation with villagers and other agencies. They also work with local water committees and agencies to repair and maintain piped water supplies.

The village heads of Mae Hae Nua, Pa Kia Noi and Huay Hoi villages have important roles in water management. The formal instruments they use include the systems and laws of the Thai state, local or village regulations evolved by water users as well as village-level customs and traditions. The village headman is the first person to meet and draw up cooperation with agencies that enter to work in the area such as the RP, Watershed Unit Office (of the RFD) and upland development NGOs such as CARE. The headman also participates in meetings and liaises with the 15 Village Watershed Network (see details in section on Mae Hae Network Committee (MHNC)). Somewhat unusually, all are also members of the local 'water-use committees'.[6]

Village heads maintain an active role in resolving water conflicts within and among villages as well as in the wider network in particular where cases of sabotage and conflict flare up in the dry season. For example, the conflict over water for Huay Pulati stream between Hmong farmers from Huay Hoi and Pa Kia Nai villages resulted in one group from Huay Hoi smashing another group's new 4-inch irrigation pipes installed in the area under management of a third group in Pa Kia Noi village. The three village headmen brought together the water users from the three groups as well as the 15 village watershed network representatives. Discussions resulted in a compromise: the new water users were allowed a smaller 2-inch pipe to take a modest amount of water from a smaller tributary and not the main stream. The previous users were allowed to stay but not to increase their cropping areas or water use.

Conflicts also occur because of differences in how rights of access to land and water are interpreted. For example, even after a household moves to live in a separate village they will use their right to land and water that they enjoyed in their previous settlement, leading to tensions with both their old and new neighbours. Minor conflicts among water users at the village level, for example over PVC pipes or water tanks, may be settled directly among those involved with the headman sometimes coordinating with the other parties involved, including the Tambon Administration Organization (TAO) members. Larger dry season water conflicts usually involve more formal meetings of the village heads and water committees.

Village headmen are often the main members of local watershed committees, at least in Hmong villages. Sometimes, the headman is really the only one who consistently participates in watershed committee meetings. This is part of the differentiation in village authority, where the ritual clan leaders deal with internal matters and the headman deals with external matters.

Mae Hae Network Committee (MHNC)

Starting in the late 1990s, 'watershed networks' emerged across sub-watersheds and watersheds in the uplands covering villages and cutting across ethnic and watershed boundaries and entire districts. Given that district-level agencies in Thailand are hesitant to enter into local-level resource issues and other line agencies use technical and hierarchical approaches, these watershed networks were able to take up the governance slack in negotiating local land and water use issues among upstream and downstream resource users. The networks are also involved in resolving resource use, sharing and access at levels of the individual village and also across the respective watersheds. In general the networks were informal groupings or forums that were taking advantage of the decentralization reforms and the increased political space at the local level created by the 1997 Thai Constitution. Some of these networks also had connections to upland NGOs while others were initiated by 'development projects' of state agencies; the latter is the case for the Upper Mae Hae watershed.

In 1988, two years after the RP entered the area, Thailand's RFD established a forestry field unit called the Mae Sa Nga Watershed Unit (MSN-WU). The MSN-WU was tasked to protect the watershed against village encroachment and recover degraded forests. The field unit is headed by a forester and assisted by several forestry guards, some of whom are hired part-time from the village. The MSN-WU as the field office of the forestry department started working with the villagers who were already part of the RP's cash crop farming network of 15 Karen and Hmong villages in Upper Mae Hae. The MSN-WU encouraged and supported the formation of the villagers' Mae Hae Network Committee (MHNC) to initiate forest protection activities.

The MHNC's activities included making firebreaks, maintaining a nursery of tree saplings for reforestation, building check dams both to prevent soil erosion and for water storage, reporting on encroachment by outsiders and subsidizing the village network's monthly meetings. Later, the network became more involved in dealing with issues of land and water use at the watershed level.

Each village sends one representative to the MHNC to attend the MHNC's monthly meetings. The forestry department through the MSN-WU pays a stipend of 500 baht to each member of the MHNC for the members' food and travel expenses. The forestry department has also set up a revolving fund of 10,000 baht for each village per year that the network can use for reforestation activities; it has also donated 5000 baht for building a check dam at the Chi Klo stream, and supported reforestation activities in the upper headwater areas. The state foresters' initial aim in setting up the watershed network was to be able to work with not only with the villagers but also with key development agencies in the area such as the RP. The foresters' network building response was also an attempt to reduce the hostility that forestry field units often face to their presence as villagers are wary of being arrested for encroachment in the watershed class 1A areas and forced resettlement. The villagers in the MHNC have used the network not only to collaborate with the foresters on forest protection activities but also to respond to local concerns about water shortages and for resolving water-related conflicts (see Box 5.3).

Opportunities and challenges for the MHNC

The MHNC's role in watershed management at the network level is undertaken with a measure of external legitimacy that comes from its association with the more formal MSN-WU. But although it has the flexibility to work at the level of watersheds or sub-watershed, it is unable to grapple with lower levels of village, sub-streams and ponds.

The network provides space for village representatives to represent their issues and interests at the monthly forums. However, in many instances it is shown as lacking authority to decide on issues, which reduces its effectiveness in resolving water-use problems.

Some of the complexities faced by the MHNC in watershed management cut across villages as well as several sub-watersheds. The network's objective of forest and water conservation can clash with villager water-users' access to the resource. In general, the decisions over water use are made at the specific level where water is used, e.g. streams, sub-streams ponds, etc. Although water-use regulations also exist, the MHNC does not have the authority to make decisions at the lower levels of ponds, streams and villages. For example, the MHNC was unable to deal effectively with the water conflict in the Huay Pulati stream.

Consensus is not easy to achieve for the MHNC as a range of opposing views, including from local leaders such as the village headman or particular water users who are affected by the network regulations, can collapse negotiations and even worsen existing village and/or ethnic tensions in the network. Issues of representation in the network are decisive; a Hmong member of the network may not represent his entire village but only his particular clan in the village.

> ## Box 5.3 Activities of the MHNC
>
> - Jointly draw up work plans for land and forest use at the monthly meetings: at the meetings, villages report on progress of any assigned conservation work (like firebreaks, etc.) and land-use changes within the community
> - Discuss the forest, water and land situation both inside and outside the villages of the network members
> - Follow-up on specific water-related problems or conflicts. One example was the conflict between the Hmong in Pa Kia Nai and Karen in Huay Khamin Nai villages over the shared use of the Huay Pulati stream. The members visited the area along with the village leaders such as the headmen, members of the local government body, RP, and the MSN-WU to look at the situation of water and land use in Huay Pulati and then discussed the issue in the monthly meeting
> - Develop a set of regulations on forest conservation, which the network uses to solve land and water-use conflicts or encroachment. These regulations are not designed to provide clear-cut details but only to offer a larger framework to solve problems of water and land use. The network regulations are broad and cannot be implemented in practice. Even though the network has tried to make water-use regulations and influence water management, regulations are more in terms of public relations, requests for cooperation and awareness campaigns
> - The regulations are often based on territorial claims (that can be and are often contested) made by the village or household using its history of prior village or household settlement;[7] this enables the particular village or individual water user to make claims over specific areas of forest, farmland or access to streams
> - Specific regulations, although rarer, also exist. For example, Clause 14 of the regulations says: 'In the flatter areas, farmland must be located at a distance of at least five metres from any nearby stream in the area. For sloping lands, the farmland must be at least 7–20 metres away from nearby streams.' This regulation is intended to protect against the pollution of the smaller streams and protect water quality
> - The monthly network meetings have a mix of representatives including villagers, RP, the head of the MSN-WU and sometimes other agencies and GOs. When specific conflicts over forest or water use need to be resolved, the representatives of the parties involved also have to take part. If they cannot be present, the agenda is postponed until such time they can join and discuss the issue

Local Government in the Khun Kan Watershed

The Khun Kan watershed is within the larger Mae Khan watershed in the Upper Ping River basin and is settled by ethnic Karen, Hmong, Lisu, Thai Leu, Shan and Chinese Haw communities in 48 villages in Bor Kaew, Yang Muen and Mae Sap sub-districts in Samoeng district in Chiang Mai province. Local livelihoods include swidden farming, paddy rice, livestock raising, collection of non-timber forest products and cash crops (Kanjan and Kaewchote, 2004).

The Khun Kan watershed has been in 'preparation' over the last 20 years to be declared as a 'national park' by the forestry department (at present, the status is national reserve forest with some parts classified as Watershed class 1A). Once it is officially declared, increased restrictions may come into place on farming and local livelihoods, including the possibility of village resettlement by the National Park Act (1961) of Thailand.

In the past, the forestry department has discussed with villagers how they can accept the national park in their midst by instructing them on forestry laws, and also surveying and zoning of forest areas. Mostly these RFD activities are undertaken through local headman, *kamnans* (sub-district head) and TAO members. (For details on the TAO, see Box 5.4.)

But the differences in perception between the villagers and the foresters over the objectives and benefits of declaring the areas as a national park means the state foresters' face difficulties getting the cooperation of villagers and their leaders such as the headman. In particular, through the TAO president of the Borkaew sub-district as well as the headman of Borkaew village, there have been requests to the RFD (so far not fully acted upon) that prior to the declaration of the national park, there should be zoning of areas for houses, farming and forest use for villagers.

Box 5.4 Local government in Thailand

As a result of Thailand's decentralization policies in the early 1990s, the Tambon Administration Organization (TAO) was established as the local government agency under the Tambon Council, based on the Tambon Administrative Organization Act of 1994. The elected village representatives of the TAO are responsible for overall development within a *tambon* (sub-district) territory including infrastructure, public health, education, social affairs and natural resource management. The revenue used for administrative costs comes from two main sources: government grants and local taxation and tariffs. Their authorities overlap with the functional mandates of other government agencies. For instance, the role of natural resource management, particularly of the forests, overlaps with the RFD's responsibility. Furthermore, the lack of specialized personnel also limits the ability of the TAOs to perform certain management roles.

The TAO Council comprises two members that are elected from each village and given a four-year term. A chairman and a vice-chairman are chosen to lead the council from the elected members. The leaders of the council are given a two-year term. The chairman and vice-chairman are not allowed to be included in TAO Administrative Committee. The TAO Administrative Committee (TAO-AC) is appointed by the District Chief. The TAO-AC comprises one sub-district leader (*kamnan*), one village leader (headman) and up to four TAO members from the TAO Council. Among the TAO members, one person will be chosen as a chairman. All members are assigned a four-year term. Unlike the TAO council and TAO-AC members, the staff of the Secretarial Office are all local government employees operating under the TAO secretary, who will also function as a secretary of the TAO-AC. This office performs several functions including treasury, public engineering and tax collection. In some TAOs, an advisory board is established, with local prominent figures invited to join as volunteers.

In general, the TAO's role is to support village development as well as natural resource management. Since the membership of TAOs comes from the local level, they are expected to have a closer understanding of local needs (in contrast to the forestry or other officials who maybe transferred every 2–3 years). The TAO is involved in building local infrastructure such as roads. But since the Khun Kan watershed area is awaiting declaration as a national park, larger concrete roads cannot be built. The roads have to be small scale, cost no more than 500,000 baht and be no more than 3 metres wide. The TAO budget also supports activities for nursery or kindergarten centres as well as rituals and traditional ceremonies such as ethnic festivals.

Tawee Damrongkiripai, Deputy President of the TAO in Borkaew sub-district, is responsible for health and environmental matters including dealing with emergency situations such as floods and forest fires. He stated that:

> *Watershed management is a process that emerges from, and of, the people and is not solely decided by officials. There are difficulties with the proposed declaration of the national park, but we would like also to see the rights of the local inhabitants protected. As a local government representative, I want the national park issue to be discussed within the TAO. When the foresters surveyed the areas and laid down the markers for the national park boundary, the TAO was not invited but there should have had more TAO involvement.*

The TAOs want more clarity about which areas would be conservation forest and which areas could be zoned off for village use, including farmlands and local forest areas. Tawee urged the handing over if possible of land titles at least for paddy lands as a way of safeguarding the local farming livelihood security (interview with Tawee Damrongkiripai, Deputy President of the TAO in Borkaew sub-district, 10 March 2009).

Meanwhile, the conundrum for the TAO is that they have to continue to proceed with local development activities: the village still needs roads and electricity even as they await the formal declaration of the national park that could lead to prohibitions on use of land areas. When the Electricity Generating Authority of Thailand (EGAT) came to set up power connections, the forestry department refused to cooperate with them. As a result, the electricity officials could not visit the village areas. In Bor Kaew sub-district, more than 10 villages do not have electricity supply but use solar cells provided by the Department of Energy. The situation with the building of schools and roads is the same. Since the area is already in a conservation zone and is proposed to be declared as a national park, EGAT is also reluctant to make investment in electricity poles and wires to provide for smaller size villages, such as those with less than 10 families (interview with Kamnan Phuka Noikaecher, Bor Kaew, 10 March 2009).

In the past, some TAOs such as in Bor Kaew have coordinated with the office of the national park in the area. Mala, the President of the Bor Kaew TAO said: 'If you declare the national park, we need to have participation of every sector including villagers, headmen, TAOs, district officials and the forestry officials.' Mala wants to have joint drawing up of plans to zone areas for village settlement and farming and use of forests.

Each of the three TAOs wants to draw up a plan that should be passed as a '*Kor Bankap Or Bor Tor*' meaning as a resolution voted on by the TAO. The villagers, including headmen and local leaders as well as foresters, would then discuss these plans further. Afterwards, once the villagers are satisfied about the zoning, and foresters accept the zoning plans, the plan will be sent to RFD and the national park can be declared. Mala says, 'The national park should be done in a way where the villagers also feel a sense of ownership of the watershed area' (personal interview, April 2008).

At present, TAOs and local leaders are collaborating between the three sub-districts and developing a network among the villages in the area about the zoning plans to build their bargaining and negotiating power with the RFD. So far no zoning plan has been drafted. Even though the national park has not yet been declared, the RFD is continuing with its zoning of Huay Taow village in Bor Kaew sub-district and mapping the areas being farmed by villagers, including both paddy rice and swidden, to exclude these farming areas from the national park. But the villagers are saying the mapping is too restrictive and must include some additional areas around these farmlands. In Huay Taow, the RFD has also taken over some area of swidden fallow that was used two years ago. The TAO has told the RFD that the fallow areas cannot be taken over since the zoning has yet to be completed. At the same time, the TAO is also telling the villagers in Huay Taow not to clear older fallow areas but continue farming in the plots that have been used in the last 3–5 years.

The RFD's view, as explained by Niwat Hongphan, head of the Khun Kan National Park, is that although the preparation of the park has been going on for the last 16 years the area is already under the Mae Khan watershed class 1A (declared in 1996). Niwat agreed that there are more than 100 communities in the area and as he does not wish to impact on their farming the officials would try to zone these areas out. He wants to assure the villagers that foresters go to each village and survey the area with the villagers, including informing them of the survey dates. The survey of the forest excludes the farms and homes as well as community forest areas. So within the conserved forest with the original demarcated area of 440,000 *rai*,[8] only 120,000 *rai* is left; they have given away up to 320,000 *rai* to the communities.

The RFD also coordinates with other relevant agencies, including the irrigation department and provincial education and health agencies, to locate ponds inside the national park. Niwat says the tensions about the national park come from a misunderstanding of the roles of the headman, TAO, RFD, etc. and that, 'Every Thai citizen has to conserve the forest not for individual benefit but to save the forests for the nation.'

DISCUSSION AND CONCLUSIONS

Using and adapting traditional and individual roles

For the past 100 years or more, the traditional leaders such as the *hee kho* and *hau zos* elders have embodied the social institutions of the Karen and Hmong communities respectively, critical to community life and management of natural resources, especially water.

In the Upper Mae Hae watershed, these traditional elders occupy the main role in water rituals in the watershed. Apart from ceremonies in the paddy fields, they may carry out rituals to request rains in very dry years, or for water spirits before opening new water supplies from streams to supply drinking water to the village. Taboos, beliefs, proverbs and songs are the main instruments used by *hee kho* and *hau zos* to manage water within the village.

The traditional leaders focus on water rituals and ceremonies as it is their customary role but, due to their experience, the fact that they are usually well-respected and their knowledge of village history, they are willingly drawn into resolving resource use conflicts, generally working together with the headman. This cooperation between traditional leaders, who continue to play a significant role in village-level watershed politics, and formal leaders like the headman has been an important facet in watershed use and management in Mae Hae over the past three decades.

Coordination with relevant upland development and state agencies is left to the formal leaders like the headman. The headmen mostly use mechanisms that are a mix of customary rules (including regulations evolved by the watershed network or the water committee) and official state laws using discretion and flexibility in interpretation and use.

Individual leaders, be they traditional elders or village headmen with the ability to be respected in negotiation, give advice, encourage compromise or work effectively at different levels or in multiple networks and play a large role in watershed management. All these three actors work primarily at the local level but have interactions at other levels. The traditional leaders and the headmen depend on and trust each other to have an impact on effective watershed management that in their case means ensuring water supplies for domestic and irrigation use, keeping tensions from flaring up to overt or violent conflicts and resolving conflicts as quickly and as much as possible within the village level without resorting to outside formal interventions such as the district authorities or police.

Upland watershed governance

Upland watersheds provide multiple environmental services to various groups – timber for logging (legal and illegal), water supply for upland and lowland rice and cash crop farming, resorts for the urban tourist, areas for orchards and fruit trees. Upland watersheds are also where state officials exert power though classifying protected areas and often undertaking 'reforestation', actions that affect the livelihood activities of upland, especially ethnic or indigenous, communities.

The differences between state and the local definitions are often reflected in divergent and competing discourses on management of upland watersheds. As Laungaramsri (1999) explains, watershed as a structuring, albeit ambiguous, concept is used by actors such as the RFD to constrain local livelihood and management efforts. At the same time, however, as this chapter has shown, formal definitions are also being reshaped by the rules of, and relationships between, local actors and institutions at different levels. And in turn, state discourses and policy influence local-level watershed management; an example was the support of village watershed networks by the forestry agency's watershed field units. By adopting and reshaping discourses, practices and tools, many groups are setting their own definitions of 'watershed' based on their cultural or customary values or transforming village-level institutions for watershed management (Wanitpradit, 2005) to retain control over their livelihoods and upland landscapes (Lebel et al, 2008).

Upland farmers are involved in a range of local institutions to collaborate among themselves as well as with the state and external development agencies and negotiate forest and water use particularly to deal with seasonal resource shortages and scarcity. At the same time, local leadership roles are both persisting (e.g. the traditional leaders) or changing (e.g. the village headman) and this in turn, influences the local politics of watershed management.

This chapter has attempted to describe the ways in which different actors and a mix of institutions or players, especially in leadership positions, are responding to the challenges of watershed management to seek effective and equitable responses and resolve resource conflicts among those who live in and use these areas for their livelihoods.

In Mae Hae, farmers are dealing with irrigation water shortages by cooperating with the RFD's watershed field units on the foresters' agenda of reforestation and managing firebreaks, while also incorporating local discussions to manage limited water resources (Ratanawilailak et al, 2008). In Khun Kan, local government bodies are taking an active role in trying to protect the development benefits as well as rights of farmers, while also trying to accommodate state laws of zoning and classification of upland areas. Within the interconnected diversity of actors and formal–informal institutions, individual roles and decisions as well as perspectives also matter. This chapter illustrates how diverse local institutions look like and act at the ground level and how these are also linked directly to individuals that make decisions and push the processes of change

The tensions between different local and formal meanings of watershed as well as management perspectives and practices continue. Where the meanings of watershed comprise a diversity of definitions and uses at the local level, or if enabling environment for local initiatives and institutional collaboration is given priority by state officials, upland watershed governance can be inclusive of upland farming livelihoods, help address resource access and go some way to resolve conflicts and tensions. The processes of compromise, negation and negotiation, as well as the reshaping and interaction of discourses and definitions that can be drawn from these case studies in northern Thailand, could help inform efforts to improve upper tributary watershed governance in the Mekong region.

Acknowledgements

This chapter draws largely from the research undertaken by Songphonsak Ratanawilailak under the M-POWER fellowship grant. We thank Dr Louis Lebel for his advice and guidance in the writing of this chapter.

Notes

1 Thailand's efforts at 'integration' of its ethnic peoples has been a dual process of incomplete integration of economic and political 'position'. There are real contradictions in these partial 'integrations' that upland people feel is a major source of political disempowerment. The upland communities are integrated politically (i.e. required to participate by upholding the legal restrictions on forest use) but not fully (i.e. not allowed to participate in any of the decision-making) and economically (i.e. required to grow cash crops to replace opium). Moreover, the upland farmers are often blamed for the ecological impacts in the lowlands (such as pollution of waterways) of these new production systems.
2 A critical event was the floods that washed down sheets of mud and cut logs down deforested hill slopes in southern Thailand in November 1988. The floods resulted in the deaths of 350 people and damage of US$120 million in property losses. The widespread public outcry over deforestation resulted in Thailand declaring a nation-wide ban on logging concessions in January 1989. This was followed in 1992 by the government revising its 1985 national forestry policy and designating 25 per cent of the total land area of Thailand as protected forest areas and aiming to expand conservation areas including national parks, wildlife sanctuaries and watershed 1A areas (Daniel, 2005).
3 In 1985, the first formal National Forest Policy was announced dividing the 40 per cent of forestland into economic or production forests and conservation or protected forests including watershed. It divided the 40 per cent of land under forests into 25 per cent for economic forests and 15 per cent for conservation forests. Subsequently, a Thai cabinet resolution declared the National Watershed Classification throughout the country. Under this classification, all major catchments were to be zoned into six categories, class 1A, 1B, 2, 3, 4 and 5, by parameters of topography, slope and elevation, type of soil, rock and forest. Class 1A, the highland area, which is considered the most sensitive part of the watershed area and is believed to be the source of water of the lowland basin, was declared a protected area in which no types of development or agriculture activity is allowed. This, however, does not include mining.

4 The term Karen refers to the subgroup Pga-k'nyaw. In Thailand the Karen are divided into four major subgroups: the Sgaw Karen who call themselves and other related subgroups Pga-K'nyaw, the Pwo Karen or Plong, the Pa-O or Taungthu who are also known as Black Karen, and the Bwe or Kayah or Red Karen. About 70 per cent of the Karen are Sgaw Karen, about 343,000 people in Thailand.
5 The poppy appears to have been an ideal cash crop to grow in the uplands of Southeast Asia. As Renard (2001, p3) says: 'Poppy grew well in the hills despite the poor tropical soils there. It required no advanced production technology nor did it need agricultural inputs such as chemical fertilizers or pesticides…Furthermore, opium as a crop has advantages in terms of marketing and handling. No cold storage or sophisticated protection against spoilage was required.'
6 Village-level water committees are made up from water users. They are chosen through simple elections. The water committee of Mae Hae Nua, for example, collects water-use fees and issues receipts, calls for village water user meetings, and sets dates and tasks for the annual maintenance and repairing of the village water supply pipes and ponds.
7 Water users (depending on their ethnic group) put forward discourses based on history of access, territorial claims, wise resource use, efficiency in irrigation, etc. to enhance their respective claims, access or rights to irrigation water. These discourses are level-dependent and act as mechanisms for negotiating access to water or resolving water conflicts.
8 1 hectare = 6.25 *rai*.

REFERENCES

Badenoch, N. and Wanitpradit, A. (2006) 'Water and institutions in the uplands of northern Thailand', USER Briefing BN-2006-08, Unit for Social and Environmental Research, Chiang Mai

Chankaew K. and Team Ae (1996) *'Kaan Kamnod Chan Kunnapaap Lumnam Korng Prathet Thai'*, Watershed Classification in Thailand (in Thai), Kasetsart University, Bangkok

Daniel, R. (2005) *After the Logging Ban: The Politics of Forest Management in Thailand*, Foundation for Ecological Recovery (FER), Bangkok

Daniel, R. and Lebel, L. (2006) 'Land policy, tenure and use: Institutional interplay at the rural-forest interface in Thailand', USER Working Paper WP-2006-1, Unit for Social and Environmental Research, Chiang Mai University, Chiang Mai

Flaherty, M. and Jengjalern, A. (1995) 'Differences in assessments of forest adequacy among women in northern Thailand', *The Journal of Developing Areas*, vol 29, pp237–254

Forsyth, T. (1996) 'Science, myth and knowledge: Testing Himalayan environmental degradation in Thailand', *Geoforum*, vol 27, pp275–292

Forsyth, T. (1998) 'Mountain myths revisited: Integrating natural and social environmental science', *Mountain Research and Development*, vol 18, pp126–139

Forsyth, T. and Walker, A. (2008) *Forest Guardians, Forest Destroyers: The Politics of Environmental Knowledge in Northern Thailand*, University of Washington Press, Seattle, WA

Ganjanapan, A. (1997) 'The politics of environment in northern Thailand: Ethnicity and highland development programs', in P. Hirsch (ed.) *Seeing the Forests for Trees: Environment and Environmentalism in Thailand*, Silkworm Books, Chiang Mai, pp202–222

Ganjanapan, A. (1998) 'The politics of conservation and the complexity of local control of forests in the northern Thai highlands', *Mountain Research and Development*, vol 18, no 1, pp71–82

Hirsch, P. and Lohmann, L. (1989) 'Contemporary politics of environment in Thailand', *Asian Survey*, vol 29, pp439–451

Kanjan, C. and Kaewchote, J. (2004) 'Communities for watershed protection: Mae Khan', in M. Thailand Poffenberger and K. Smith-Hanssen (eds) *Asia Forest Network*, Asia Forest Network, Bohol, The Philippines

Knie, C. and Moller, K. (1999) 'Watershed Classification with GIS as an Instrument of Conflict Management in Tropical Highlands of the Lower Mekong Basin', Paper presented at the GTZ workshop on Application of Resource Information Technologies (GIS/GPS/RS), in *Forest Land and Resource Management*, 18–20 October, pp146–154

Lebel, L., Daniel, R., Badenoch, N., Garden, P. and Imamura, M. (2008) 'A multi-level perspective on conserving with communities: Experiences from upper tributary watersheds in montane mainland southeast Asia', *International Journal of the Commons*, vol 1, pp127–154

Laungaramsri, P. (1999) 'The ambiguity of watershed: The politics of people and conservation in northern Thailand. A case study of the Chom Thong conflict', in M. Colchester and C. Erni (eds) 'Indigenous Peoples and Protected Areas in South and Southeast Asia: From Principles to Practice', IWGIA Document Number 97, International Working Group for Indigenous Affairs, Copenhagen

Laungaramsri, P. (2001) *Redefining nature: Karen Ecological Knowledge and the Challenge to the Modern Conservation Paradigm*, Earthworm Books, Chennai

Pratong, K. and Thomas, D. (1990) 'Evolving forest systems in Thailand', in M. Poffenberger (ed.) *Keepers of the Forest: Land Management Alternatives in Southeast Asia*, Kumerian Press, West Hartford, CT

Puginier, O. (2002) '"Participation" in a conflicting policy framework: Lessons learned in a Thai experience', ASEAN Biodiversity Special Report, vol 2, no 1, pp35–42, ASEAN Centre for Biodiversity Conservation (ARBC), Los Banos, The Philippines

Ratanawilailak, S., Daniel, R. and Lebel, L. (2008) 'Local institutions, strategies and discourses in dry-season water management: Upper Mae Hae watershed in northern Thailand', USER Working Paper, Unit for Social and Environmental Research, Chiang Mai University, Chiang Mai

Renard, R. (2001) *Opium Reduction in Thailand 1970–2000: A 30 Year Journey*, Silkworm Books, Chiang Mai

Suraswadi, P., Thomas, D.E., Pragtong, K., Preechapanya, P. and Weyerhaeuser, H. (2005) 'Northern Thailand: Changing smallholder land use patterns', in C.A. Palm, S.A. Vosti, P.A. Sanchez and P.J. Ericksen (eds) *Slash-and-Burn Agriculture – The Search for Alternatives*, Columbia University Press, New York, pp355–384

Tangtham, N. (1992) 'Principles and application of watershed classification in Thailand', Training Course on Natural Resource Management and Conservation in Watershed Areas, Asian Institute of Technology, Bangkok

Thomas, D. (2002) 'Landscape agroforestry in northern Thailand', ICRAF, Chiang Mai
Thomas, D. (2005) 'Participatory watershed management in upper northern region of Thailand', ICRAF, Chiang Mai
Thomas, D.E., Preechapanya, P. and Saipothong, P. (2004) 'Developing science-based tools for participatory watershed management in montane mainland Southeast Asia: Final report to the Rockefeller Foundation', ICRAF, Chiang Mai
Vandergeest, P. and Peluso, N.L. (1995) 'Territorialization and state power in Thailand', *Theory and Society*, vol 24, pp385–426
Wanitpradit, A. (2005) 'Dynamic of local knowledge and alternative management of resources on the highlands: A case study of Hmong Mae Sa Mai community in Mae Rim district', Chiang Mai University, Chiang Mai
Wanitpradit, A. (2008) 'Emerging local institutions for watershed governance in the ethnically diverse Mae Tian sub-watershed, Mae Wang District, Northern Thailand', M-POWER Fellowship Report, Unit for Social and Environmental Research, Chiang Mai
WCC (1983) 'Study on watershed classification of important river basins in Thailand', report presented to the First Workshop on Watershed Classification in Thailand, Chiang Mai, 16–18 December

6

Gender, Commercialization and the Fisheries–Aquaculture Divide in the Mekong Region

Louis Lebel, Santita Ganjanapan, Phimphakan Lebel, Mith Somountha, Tran Tri Ngoc Trinh, Geeta Bhatrai Bastakoti and Chanagun Chitmanat

INTRODUCTION

Inland fisheries and aquaculture are crucial to the well-being, livelihoods and subsistence of many men and women living in the Mekong region (van Zalinge et al, 2001; Hortle and Bush, 2003; Baran et al, 2007). While inland capture fisheries may be becoming less resilient to over-exploitation and other environmental changes (Dudgeon, 2005), their value and significance have often been underestimated or ignored by policy-makers (Friend, 2009). Claims about poor condition of, and weak prospects for, capture fisheries have been common justifications for promotion of aquaculture (Bush, 2008; Friend et al, 2009). An ideological divide has emerged between those actors who believe that the emergence of an aquaculture industry is part of the solution and others who see it as part of the problem, in which commercialization leads to environmental degradation (Bene, 2005; Bush, 2008). In part, this dichotomy has been created by coalitions with different interests competing for policy attention and investments in sector development and capacity building. Regardless, aquaculture has grown in importance across the Mekong region and, like wild capture fisheries, has become increasingly industrialized and commercialized (Lebel et al, 2002; Phillips, 2002).

Women engage in a diverse range of aquaculture- and fisheries-related livelihoods. Recognition and rewards are often incommensurate with labour contributions, raising questions of fairness and needs for empowerment (Kelkar, 2001). A persistent belief in rural gender studies is that commercialization is bad for women (Wilson, 1985). Evidence from studies of agricultural transformation in developing countries is mixed, reflecting variable patterns in social organization, property rights and the importance of interactions with class and ethnicity. Women can benefit from engagement with the market, continue to control land, technology and income associated with new export crops (Hamilton et al, 2001). Commercialization can result in changes in household structure with women and older members increasingly doing farm work (Pingali, 1997).

A few studies have concluded that commercial fish aquaculture has negative implications for women (Galmiche-Tejeda and Townsend, 2006). Kusakabe (2003) introducing a study of fish farming in northeast Thailand writes, 'It has now become widely accepted that the more aquaculture activities are commercialized, the less women are involved.'

This chapter reviews evidence from the Mekong region about how gender relations affect, and are affected by, commercialization and transitions from capture fisheries to aquaculture. Several states and transitions can be distinguished (see Figure 6.1). Livelihoods may be largely oriented around capture (A, B) or culture (C, D) of fish and production subsistence (A, D) or commercial-oriented (B, C). Commercial aquaculture (C), for example, may arise out of agriculture (E) or subsistence fishing (A). Transitions may involve complete replacement of livelihoods or be partial, as when, aquaculture becomes a supplementary livelihood in otherwise fishing oriented livelihood.

We hypothesize that several dimensions of gender relations are likely to be important to one or both of these transformations. Access to different resources and livelihood options reflect how work is divided between men and women. Contributing labour may be critical to staking claims to productive outputs. On the other hand if labour contributions are under-recognized they may also lead to a multiplication of roles and burdens. Being engaged in an activity may or may not translate into having an influence on decisions about investments of time and funds or in managing shared resources (see Figure 6.1).

DESIGN

To explore the effects of these transitions on gender relations we undertook a systematic review (Dixon-Woods et al, 2006) of studies about gender and aquaculture or fisheries in the Mekong region. For the purpose of this study the Mekong region is defined as the countries of Burma/Myanmar, Cambodia, Lao PDR, Thailand, Vietnam and Yunnan province in China. We included studies of inland and coastal, reservoir, pond and river-based aquaculture of fish or other aquatic animals.

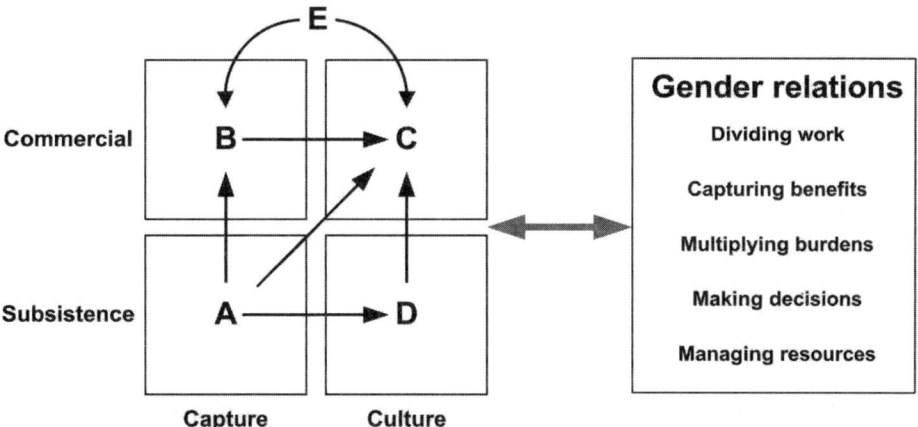

Figure 6.1 *Transitions from capture to culture or commercialization may affect gender relations and vice versa*

A: subsistence fisheries
B: commercial capture fisheries
C: commercial aquaculture
D: subsistence aquaculture
E: agriculture

Source: Authors

We searched Academic Search Premier EBSCO (www.ebscohost.com), JSTOR (www.jstor.org) and Web of Science (www.isiwebofknowledge.com) and GOOGLE Scholar (http://scholar.google.com), using, 'aquaculture' or 'fisheries', and 'gender' or 'women' as key terms in combination with country names. A few key and substantive agency reports were recommended to us by research colleagues working on fisheries in the Mekong region and also included in the review. Different articles or reports about the same area by same authors or organization were considered a single study. This procedure led to the identification of a set of 19 core studies (see Table 6.1). Core studies presented empirical research, described their main methods and were explicitly concerned with issues of gender in inland fisheries or aquaculture in the Mekong region. Other studies that met only some of the criteria provided additional counter or supporting evidence for wider generalizations. The core studies were geographically distributed as follows: Cambodia (7); China (2); Laos (2); Thailand (5); Vietnam (3).

The rest of this chapter is organized around the five dimensions of gender relations (see Figure 6.1): dividing work, capturing benefits, bearing burdens, making decisions and managing resources.

Table 6.1 *Core studies on gender in aquaculture and fisheries from the Mekong region*

Location	Physical setting	Production system	Transition type	Key references
Prey Veng and Svey Rieng provinces, Cambodia	Ponds	Semi-intensive, small scale	A=>C,D	So et al, 1998; Nandeesha et al, 1996
Kampong Chhnang province, Cambodia	Lake	Capture fisheries and aquaculture	A=>B,C	Mith, 2008
Kampong Chhnang province, Cambodia	Lake	Capture fisheries	A=>B	Gätke, 2008
Various provinces around Tonle Sap Lake, Cambodia	Lake	Capture fisheries	A=>B	Resurreccion, 2006; 2008
Several provinces in coastal, Mekong and Tonle Sap, Cambodia	Lake, river and coast	Capture fisheries	A=>B	MAFF and CBNRM, 2008
Several provinces around Tonle Sap, Cambodia	Lake	Capture fisheries	A=>B,C,D	IFM and ADB, 2007
Aranyaprathet and Poipet on the border of Thailand and Cambodia	Markets	Trade in products from capture fisheries and aquaculture	A,E=>C	Kusakabe, 2004; Kusakabe et al, 2006
Lijiang County, Yunnan province, China	Reservoir	Capture fisheries in a dam reservoir	E=>B	Yu, 2001
Lijiang County, Yunnan province, China	Pond	Aquaculture	E=>C	He, 2001
Savannakhet province, Laos	Lake	Stock enhancement of natural water body	A,B=>C	Garaway, 2006
Several provinces in Lao PDR	Pond	Semi-intensive aquaculture	E=>C	Murray et al, 1998
Upper Ping River, Chiang Mai and Lamphun provinces, northern Thailand	River	Intensified aquaculture	E=>C	Lebel, 2008; Lebel et al, 2008; 2010
Udon Thani province, northeast Thailand	Pond	Semi-intensive and intensified aquaculture	D=>C	Kusakabe, 2003
Maha Sarakham province, northeast Thailand	Pond	Semi-intensive and intensified aquaculture	D=>C	Sullivan, 2006

Table 6.1 *(Continued)*

Location	Physical setting	Production system	Transition type	Key references
Buriram province, northeast Thailand	Pond	Integrated agriculture-aquaculture	E=>C	Setboonsarng and Edwards, 1998; Setboonsarng, 2002
Lower Songkhram River Basin, Sakhon Nakhon and Nakhon Phanom provinces, northeast Thailand	Wetland	Fishing and cage culture	A,B=>C	Sriputinibondh et al, 2005
Nam Dinh and Binh Dinh provinces, central Vietnam	Coastal pond and mudflats	Semi-intensive aquaculture (shrimp, clams)	A=>C	Le, 2006; 2008a; 2008b
Tam Giang Lagoon, Thua Thien Hue central Vietnam	Lagoon	Cage aquaculture	A=>C,D	Suong, 2006; Nhung, 2008
Son La, Lai Chau and Hoa Binh provinces, north Vietnam	Pond	Integrated aquaculture	E=>C	Kibria and Mowla, 2006

Transition types are as shown in Figure 6.1

DIVIDING WORK

Opportunities to engage in specific livelihood activities often differ for men and women for a variety of reasons. Even when basic access to required resources is open, work is usually divided in ways that reflect gender norms of particular cultures – national, ethnic and organizational (Agarwal, 1997; Li, 1998). Division of labour is one of the fundamental and better-studied dimensions of gender differences and relations in capture fisheries and aquaculture (Kelkar, 2001; Sharma, 2004). In this section we look first at capture fisheries, then at aquaculture and end with interactions between the two and with other parts of the commodity chain such as trading.

Capture fisheries

There is substantial variation across the Mekong region in extent and roles in which women and men are involved in capture fisheries (Bourdier, 1998; Siason et al, 2002; Garaway, 2006). Women engage in a variety of fishery-related activities including small-scale capture, cleaning and fixing fishing gear and collecting aquatic plants and animals (MAFF and CBNRM, 2008). Women are often most prominent in processing and trade (Kusakabe, 2004). This variation can be partly explained by access, which, in turn, can depend on capabilities, rights, social ties and other institutions.

Access and ability to exploit fishery resources in the Tonle Sap Lake in Cambodia is strongly dependent on size of operations, despite fisheries laws that are supposed to help small-scale fishers. Small and medium-scale fishing gear that are used by small and medium-scale fishers cannot compete with large-scale intensive fishing gear such as long *oun* (seine net) that is pushed by two big engine-driven boats on both sides of the net. Small and medium-scale fishers find it difficult to catch a sufficient amount of fish compared to those with the most advanced technology, the *oun* operators. The *oun* operators can catch 100 times more fish than small-scale fishers (Mith, 2008). Power dynamics means that women, as users of smaller-scale fishing gear, continue to be at a disadvantage in practice under the fishing rules (Dowley, 2008).

Community fisheries groups were created as part of reforms that released a fraction of fishing concession lots in Tonle Sap Lake for community management. Women's involvement in community fisheries is low; women's work in land agriculture, it is suggested, subsidizes fishing and fisheries management activities of men (Resurreccion, 2006).

In Resurreccion's (2006) view male-dominated patron–client relationships with military and political links continue to dominate distribution of access rights in Cambodia's fisheries. Greater involvement of women, given current social institutions and gendered practices, is likely to increase the burden of women with few corresponding benefits. Social circumstances matter greatly to the empowerment impacts of participation in community management initiatives. Already vulnerable, female-headed households struggle for legitimacy and thus to gain access to needed resources under new management institutions. Those who do not engage in community fisheries are poorer and forced into more marginal areas of the lake to meet basic subsistence needs (Resurreccion, 2008).

Surveys of fishery dependent communities in several provinces in Cambodia in the mid 1990s reported 17–29 per cent of women-headed households (IFM and ADB, 2007). Such households had, on average, fewer assets, limited access to land and poor quality housing. Surveys in 2003 in Siem Reap and Kampong Chhnang showed that fishing, aquaculture and fish processing were important occupations for women household heads. Overall women contributed to about 20 per cent of fishing effort, and were even more involved in the maintenance of equipment and supply of gear. Some studies suggest that women's participation in fishing may be increasing as catches fall and catch efforts are increased in an effort to maintain production (MAFF and CBNRM, 2008).

Access to fishery resources in Chnok Trou village can be seen through power relations, social relations and capital investments. These factors have an influence on the division of work between men and women who have to adapt with to new conditions in order to sustain household livelihoods (Mith, 2008). The change in the gender division of work also has an effect on gendered space as well. Gendered space can be changed with respect to the change in gender division of work and interests. The spaces of both women and men are sometimes overlapping and interdependent on each other and differ according to age and economic status.

Women in Thailand participate in fishing local waters as well as pre- and post-fishing activities. Women help their husbands clean and repair fishing gear and are usually the ones who sell products in the local markets, drying or fermenting what is unsold. In the lower Nam Songkhram River women fish if men's labour is not available, but common perception is that the task is not suitable for women, especially on their own (Sriputinibondh et al, 2005). Women rarely participate in larger-scale, commercial fishing. Industrialization of fisheries around Thailand's Songkhla Lake suggests that women may be doing less fishing from home and working more in fish processing factories (SDF and FSF, 2005).

Division of work in capture fisheries in the Mekong region depends on things like size of operations, how rights of access are determined, alternative livelihood opportunities and power relations. At the same time there are often strong inter-dependencies between work done by men and women that may also constrain the ease with which gender relations can shift.

Aquaculture

As for capture fisheries there is substantial variation among locations in how labour is divided in aquaculture (MRC-TAB, 2006). Some of the variation can be explained by differences in wealth or social class, social ties, production technologies and how innovations are introduced.

Aquaculture innovations can be gender exclusive. A district in Nam Dinh province, central Vietnam, introduced four shrimp ponds that were then leased to groups of 5 to 10 households (Le, 2006; 2008b). Shrimp pond owners in the study claimed that 'shrimp farming is a risky business and requires large amounts of capital, so women are unable to participate' (Le, 2008b, p30). Women's only role was to collect seaweed. Clam farming was also popular in this village and their trade a source of wealth. The Village People's Committee granted farming sites on mudflats and collected rent. This process excluded poor and women-headed households who instead had to work as collectors on other people's farms (Le, 2008b). In this area both women and men traded fisheries products. Women, however, were excluded from trading farmed shrimp 'because it was more stable and generated more profits' (Le, 2008b, p36) and thus earned much less for similar work efforts than men. The patriarchal system excluded women from more profitable jobs and social practices and kept the shrimp traders' network entirely male (Le, 2006).

In a related study in Binh Dinh province similar patterns of exclusion were observed (Le, 2008a). Expansion of shrimp farms came at the expense of women and girls who used to collect aquatic products in coastal mangroves. Women were again largely excluded from shrimp farming on newly privatized land; in only 4 of 322 households headed by men were women engaged in shrimp farming activities. Even female-headed households were excluded from acquiring land that became available as part of economic and property right reforms that led to issue of 'red books' (Le, 2008a), a form of formal title that can be used as collateral for bank loans.

Fish ponds introduced into upland communities in northern Vietnam are actively worked by women, but never without support of men (Kibria and Mowla, 2006). Men, for example, did most pond preparation and stocking, while both sexes were involved in feeding, fertilizing, harvesting and marketing. Absence of men, because of work opportunities that take them away from home, may create opportunities for women to engage in local aquaculture or fisheries activities (Dang, 2008). On the other hand lack of labour can also be a constraint (Suong, 2006).

In the upper Ping River in northern Thailand, women were seen as more reliable and careful, and thus better at taking care of fish. 'Women are more responsible. Men maybe get drunk in the evening and forget to feed the fish at the right time. They know it is their job but they don't do it on time' (Lebel et al, 2010, p209). But fish farming also involves tough physical work and this is a basis for division of labour. Heavy lifting and pulling work with cages is men's work. In the upper Ping women who run farms usually employ, or get help from men for tasks that require physical strength (Lebel et al, 2010).

Citing some earlier local studies in Cambodia, Suong (2006) notes that women, overall, had less opportunity to engage in aquaculture than men because of low education and lack of physical strength, and because they were often not listed in household registration documents. In addition, women hired to fish in net enclosures were paid less for the same hard work. Diversification of income sources increased the burden on women (Suong, 2006).

Integrated farms, known locally as *aset-pasom-pasan*, combine rice, fish, vegetables, fruit trees and livestock activities around an aquaculture pond (Setboonsarng, 2002). The inclusion of aquaculture ponds into rice-dominated farming systems in Buriram Province, northeast Thailand, led to increases in amount of work of both men and women compared to non-integrated farms. Women's workloads increased more than men's: as fish culture became established on a farm women did more and more of the feeding and maintenance activities; total time allocated by both men and women also increased, suggesting greater intensification. Much of the labour for fish farms when they first started was spent collecting and preparing feed from on-farm wastes. As farms became more established farmers switched to using commercial pelleted feeds and, consequently, a less integrated system (Setboonsarng, 2002).

He (2001) studied aquaculture in a mixed Han-Naxi village in Lijiang county, Yunnan province, China. A long history of a collectively owned fish pond changed in 1983 when the village started allocating quotas to individual households to build and maintain riverside ponds harvested once a year. Over recent years all farmers have converted their ponds to service recreational (angling) fishing demands as this provides higher profits than raising young fish. Women participate actively in fish farming, collecting feed, changing water, collecting fees and serving anglers. Men guard ponds at night and often transport fish. Women also market and sell fish after the annual harvest. Women have greater decision-making power in fish farming than in many other activities; fish farming, however, also means increased workloads and more limited mobility.

Interactions and beyond

Shifts from capture to culture, or the addition of culture to land-based agricultural livelihoods, produce a range of interactions that can impact on, or be affected by, gender relations. Most understanding has been gained at household or local community levels. Outcomes in production and retailing parts of commodity chains also differ, reflecting norms and social processes that reinforce tasks as being masculine or feminine.

A study in southeastern Cambodia found major differences between contrasting provinces – Prey Veng with substantial capture fishery opportunities and Svey Rieng with much less – in levels of involvement of women and commercial orientation of aquaculture programmes initiated in the early 1990s by an NGO called PADEK (Partnership for Development in Kamphuchea) (Nandeesha et al, 1996; So et al, 1998). Division of labour was sharp with men doing most of the pre-stocking activities like digging ponds and applying lime and fertilizers. Women mostly took care of fish, feeding and fertilizing, once the pond was stocked. In Svey Rieng women often initiated and carried out most aquaculture activities and felt confident they could carry on without men (So et al, 1998). In Svey Rieng most production was consumed in the household whereas in Prey Veng more than half was sold. The researchers stated that, 'Women did not consider fish culture as an additional labour, but viewed it as a food source and recreation through daily feeding of fish and better utility of time' (So et al, 1998, p22). Extensive surveys in Cambodia suggest that women contribute about half the total labour to cage culture in the lake, are more involved in processing than men, and dominate fish marketing (IFM and ADB, 2007).

Tam Giang-Cau Hai lagoon in central Vietnam supports the livelihoods of about 300,000 people, capture fisheries and aquaculture being the main sources of income for many households (Tuan et al, 2009). Engagement with aquaculture increases household incomes. Women engage in both fishing and aquaculture (Suong, 2006). Men sleep near fish cages to guard them (Nhung, 2008). Men build and move cages, whereas women maintain cages, for example, by repairing nets. Both men and women feed, harvest and sell fish (Nhung, 2008). Women mostly work with mobile fishing gears (push net, clam collection), which is hard work but yields low income. Men did not use these technologies, an informant explaining it this way: 'This activity is very hard…Only women practice it, because you see, women are industrious by nature and want to save every amount of money' (Suong, 2006, p43).

Young women dominate assembly lines in fisheries processing factories and pond sides when fish or shrimp from aquaculture farms are sorted. Thai employers believe that women are more suited to jobs requiring concentration (Mills, 2001) and 'light hands and nimble fingers' (Lebel et al, 2010). Working conditions, job security and opportunities for promotion and skill development for women, however, are often poor (Sharma, 2004).

In the lower Nam Songkhram River wetlands in northeast Thailand, women are primarily involved in fish processing and marketing (Sriputinibondh et al, 2005). Processing includes sun-drying, smoking, salting and fermenting. While men do most of the fishing women also help when male labour is missing or insufficient. In aquaculture the division of labour is different: women now do half of the work feeding and rearing fish although men usually prepare the fish cages (Sriputinibondh et al, 2005).

In commercial settings, women may have an advantage over men in gaining entry. Men and women often work together. In Cambodia many husband-and-wife teams buy fish from fishers or lot owners (Kusakabe, 2004). In small-scale border trade women lead the activity but are often supported by male relatives. In the fish trade on the Thai–Cambodian border, for instance, most small traders are women – an extension of conventional domestic role in fish retail (Kusakabe et al, 2006). Entry into larger-scale export or wholesale businesses, however, is difficult without capital and because it requires relationships with many other actors including government officers.

Transitions from agriculture to fisheries may be by necessity rather than choice. The trans-watershed supply project for irrigation and domestic supply water for the Lijiang valley, Yunnan, China, converted the Lashi wetland into a controlled lake (Yu, 2001). After the construction of the dam the reservoir flooded significant areas of farmland. Some compensation was offered but not for all the land. Moreover normally cultivated areas became more vulnerable to high water levels after floods. In part as a coping strategy reflecting the changed ecology of the landscape households become more involved in fishing activities. Although most larger-scale fishing work is by men it was not exclusively so. Women dominated fish trading and viewed the expansion of the capture fishery as an opportunity, in part, because they usually had better knowledge of market channels than men (Yu, 2001).

In the Mekong region, women participate to a variable degree in most parts of the fisheries and aquaculture sector and typically predominate in marketing and retailing although this is changing with commercialization and industrial reorganization. Access to particular livelihoods and time spent in different tasks are often strongly influenced by social norms and gender relations. Women and men typically do different tasks and have different responsibilities. The exact mix differs from place to place and may change with time. Tasks of men and women are usually interdependent, although women's work is usually less visible.

Capturing Benefits

Capturing the benefits of engagement in new livelihood activities often depends on gaining recognition for labour inputs and taking some control over revenue streams (Goetz and Gupta, 1996; Kenney, 2006). Being able to make work contributions visible, either separately or jointly, can be critical to secure rights over property (Li, 1998). This, in turn, may help improve bargaining power within the household and lead to a larger role in decision-making within and beyond the household (Agarwal, 1997; Kabeer, 2005).

Making work visible

Women's contributions to labour in fisheries and aquaculture are larger than is normally recognized (Williams et al, 2005). One reason is that surveys of fisheries neglect the various ways women contribute, including: collecting fish and other aquatic plants and animals to meet daily food requirements of families, supporting husbands in building and repairing fishing gear, processing fish and selling products (Suntornratana and Visser, 2003). Many of these activities are seen as 'supporting' and 'non-income generating' and that can make them less visible.

Perception of a livelihood activity as primary or secondary affects how time spent pursuing those activities is valued by both women and men (Coyle and Kwong, 2000). If taking care of fish is seen as a secondary activity its contribution to family earning and well-being is made less visible. Women's activities in lower Nam Songkhram fisheries are unpaid and largely unrecognized; in aquaculture men get more credit because they do some of the physically demanding tasks (Sriputinibondh et al, 2005).

Women's status on integrated farms with fish ponds in northeast Thailand was higher than on conventional farms; their participation, access and control of resources increased (Setboonsarng, 2002). It is not entirely clear if this increase matched increases in workloads that already were higher than that of men on conventional rice farms.

In the upper Ping, adding, or switching to, fish farming was largely seen in positive terms, and better than just taking care of orchards, by both men and women. Women who farm fish express satisfaction with their work and income it brings to their families (Lebel et al, 2010). Women achieve similar profits to men. Women who farm fish for commercial profit have become knowledgeable managers and important decision-makers on investments (Lebel et al, 2007).

Women agreed that a project introducing aquaculture into farming systems in upland areas of northern Vietnam brought benefits, such as better nutrition for women and children and income to pay for schooling of children (Kibria and Mowla, 2006). Women also said that the clear division of labour and responsibilities in fish culture was beneficial to them. Improved technical knowledge, for example, gave women a better position from which to discuss investments and management practices. Engagement with the project also strengthened local women's unions. In these communities men were the primary decision-makers and asset-holders. On the other hand, tradition and religion can turn increased engagement into reduced status. For Muslim women in southern Thailand, working outside the home to repair nets or in factories brings income but also criticism from their men folk that these women are of lower, Thai-buddhist, moral standards (Dorairajoo, 2002).

When it comes to paid work, women's contributions are valued less than men's: in the upper Ping, for example, as a worker on another person's fish farm women could expect 100–120 baht/day whereas men received 150–200 baht (Lebel, 2008).

Men clearly benefit from continuation of existing arrangements that subordinate women and in everyday actions and discourse women often reconfirm their subordinate status (Coyle and Kwong, 2000). Women in Cambodia are expected to be subservient in public interactions. Women and men make use of this norm. In border trade women are better able to negotiate with officials to reduce fees; men are reluctant to be in such a position and more likely to get into fights (Kusakabe et al, 2006). On the other hand women as small traders remain vulnerable to exploitation and arbitrary rulings.

Controlling receipts

Women seek supplementary and independent incomes that help them both in meeting household level food security goals as well as improved financial independence and control as individuals (Mills, 1997). But, getting a loan, engaging in a productive activity, earning money from it and being in control of that activity and of the income it produces, are very different matters (Kenney, 2006). Improved credit access, for example, does not automatically translate into transformation of gender relations: control of loans and income is critical (Goetz and Gupta, 1996).

Women in Tonle Sap fishing communities have primary responsibility for financial management often taking formal and informal loans to meet short-term food security needs (Resurreccion, 2006; Mith, 2008). Women typically trade fish and are able to get fishing gear materials on credit from local shopkeepers with installment payments in the form of households' daily fish catch (Mith, 2008). Women have taken on these activities because they have made good social relations with traders. Having better social networks with outsiders than men gives them stronger bargaining power within the household (Mith, 2008).

Another study in southern Laos likewise found that women sold fish and controlled the receipts from sales (Noraseng et al, 1999). Cultural differences, associated with ethnicity, are important in Laos. Hmong men, for instance control cash, so they take responsibility for buying fingerlings whereas in Lao Loum households women could make the purchases (Murray et al, 1998). Lao Loum women also control receipts from sale of fish and use it to purchase household items.

In rural Thailand women have traditionally had major responsibilities in marketing and financing. Women are better at 'making sure there is enough food to eat'. In fish farming households in the upper Ping, a range of financial arrangements was observed (Lebel et al, 2010). Some pooled their incomes whereas others maintained separate bank accounts: 17 women from 33 couples had their own bank accounts, whereas in 13 cases only the husband had an account.

Somewhat unusually in integrated farms in Buriram (northeast Thailand), men were largely responsible for fish marketing, as in this location most sales were made directly at the pond edge and men did most of the harvesting (Setboonsarng, 2002). When women sold fish, they usually kept the money and therefore control over benefits of fish farming (Setboonsarng, 2002).

Community fisheries as part of larger rural development packages may create other opportunities for women to secure benefits. Resurreccion (2008) argued that women who get involved in community fisheries activities often have husbands or relatives in leadership positions. A key motivation to engage was access to low interest loans (Resurreccion, 2008) or related opportunities for livelihood improvements (MAFF and CBNRM, 2008).

Sullivan (2006) in another study from northeast Thailand concluded that commercial 'cage culture appears to bring significantly more benefits to men who can more readily access and utilize the benefits of increased incomes and asset ownership'. She drew this conclusion indirectly based on observation that engagement in cage culture did not improve the mobility of women beyond the local community whereas for men it did. An alternative explanation is that gendered norms of mobility are hard to change and a shift to aquaculture is not enough to do so, although it does bring in other benefits.

The capacity of women to capture benefits from their own and families' engagement in fisheries or aquaculture is highly context-sensitive. Cultural and inter-household differences in recognition of labour inputs, responsibilities for financial management and gender-sensitive expectations all play a role. New activities may provide opportunities for new partitioning of benefits but only if intransigent gender norms are sidestepped or can be shifted and it is clear from studies in the Mekong region that this is often difficult.

Multiplying Burdens

One of the consequences of women's work being less visible than that of men is that diversification of livelihoods often multiplies burdens. Women usually have greater responsibilities for a range of time-consuming tasks in and around the home (Coyle and Kwong, 2000; Power, 2004). Post-harvest processing of capture fisheries products and aquaculture are often introduced, at least initially, as a supplemental livelihood option contributing to the diversification strategies of rural households. But little attention is given to real time available and existing workloads with women usually being expected to cope (Kelkar, 2001; Williams et al, 2005).

Working at home

The household or family is an important social unit, constraining and supporting how women and men engage with old and new livelihood opportunities and markets (Wilson, 1985). How women perceive relationships within the family institution is crucial to everyday practices (Agarwal, 1997). Socialization begins 'from a young age, the girl is taught to stay at home, take care of the family, and be responsible for the housework' (Jongudomkarn and West, 2004). Marriage, family and religious institutions have interacted profoundly to shape rural and then industrial development's interactions with gender relations in the Mekong region (Mills, 1997; Coyle and Kwong, 2000; Curran et al, 2005).

Women in Tampuan villages in Ratanakiri (Cambodia) had much higher work burdens than men and as a result of poor living conditions were exposed to many diseases and in poor health (Bourdier, 1998). Suong (2006) also speculates that long hours associated with clam collection, for instance, may have health implications. In Cambodian fishing communities, men typically have significantly more time for recreation and sleep than women (MAFF and CBNRM, 2008).

Caring for children and husbands and doing housework are demanding and has an associated opportunity cost that is not taken into account in most societies' framing of 'economic activity' (Power, 2004). As noted earlier, work at home is invisible and as a consequence new activities like aquaculture or community fisheries management multiply work burdens. To cope with multiple burdens, women in Thailand lean on Buddhist teachings around self-reliance and its framing of mother's role as one of honourable sacrifice for their children's happiness (Keyes, 1984; Mills, 1997; Jongudomkarn and West, 2004). Buddhist social norms create burdens for women along with the means to cope with them. Likewise, in the Mekong delta of Vietnam women are taught from an early age to sacrifice their time and energy for parents, siblings and husbands (Taylor, 2007).

Taking care of a fish pond or set of cages must often be integrated into household chores women are expected to carry out such as looking after children, cleaning the house and taking care of elderly relatives (Sriputinibondh et al, 2005; Lebel et al, 2010). Moreover, women are expected and willing to take over extra farm duties when their husbands have left, for example, to take up employment overseas or in the city (Sullivan, 2006). Subsistence-oriented aquaculture production is considered an extension of household maintenance work and thus part of a women's work (Kusakabe, 2003).

Fish farming for many households in the upper Ping is one activity in a portfolio of income-generating activities over the year. Men are more likely to have employment in construction and to work on orchards, whereas women are more involved in retailing food or agricultural products (Lebel et al, 2010). For those with orchards, having a fish farm is compatible as they can work on trees between feeding times: 'It is a livelihood that fits into the daily cycle. I can still take care of the family and make a profit' (Lebel et al, 2010, p213).

Male fish farmers in Tam Giang Lagoon in Vietnam said similar things: 'Raising fish can be done in my free time. If we don't raise fish we have much free time. In that time, we sometimes go to rice fields to weed, or sleep, or gather together and drink wine' (Nhung, 2008, p32). In this setting even female-headed households were able to successfully raise fish, with help for physically demanding tasks from kin.

Fish and shrimp farming differ from some other rural livelihood options in their peculiar time and investment demands. Farms need attention several times a day, to clean out waste in the cages, and to feed animals, in both systems. Each round may not take very long, but it has to be done 3–5 times per day, constraining mobility. A Department of Fisheries official in the upper Ping said aquaculture is suitable for women because, 'Men often go out to work elsewhere. Women are at home and can look after cages. Fish cages don't move' (Lebel et al, 2010, p214).

Moving around

Several researchers have noted micro-positional effects on gender. Women work on fish farms more when they are close to the home garden in Laos (Murray et al, 1998). They argue this is because feeding can be fitted into their daily routines. Farms near the house are perceived as part of the 'domestic' sphere. While both men and women could buy fingerlings in Laos men were more likely to do so when distant travel was involved. Location of ponds relative to households also influenced who fed fish: if ponds were far away women were less likely to do the feeding than when ponds were near the house.

In a study in Savannakhet province (Laos) both men and women from all groups were actively engaged in fishing (Garaway, 2006). Women, however, were often constrained to water bodies closer to the village and were expected to combine fishing with other activities like child care. Women used gear that caught smaller fish, mostly for household consumption, whereas men's gear could also catch bigger fish and be sold. In community fisheries in Cambodia, armed teams of men do patrolling, often at night (Resurreccion, 2006). Women are rarely involved but when they are some observations suggest they may be more effective than men in securing compliance of others to local rules (MAFF and CBNRM, 2008).

Sullivan (2006) showed that the mobility of women beyond their immediate community was less than that of men both before and after they had engaged in fish culture. Low mobility as Sullivan rightly pointed out could matter: for example, in restricting access to technical information in aquaculture as well as more generally for using rights to education, health or political participation. Engagement with commercial aquaculture, because corporate partners bring inputs to the farm and come and collect the harvest, was consistent with pre-existing norms on mobility. Mobile phones, on the other hand, substituted for vehicle-based mobility.

Lack of mobility is a barrier to women entering wider trading networks for cultured and capture products in coastal Vietnam, especially for the poorest households that cannot afford a motorbike (Le, 2008b). Lack of mobility, however, may be compensated for by depending on kin-based networks, but this tends to reinforce the weaker position of poor women. Women headed many of the poorest households in this study.

The Cham, minority Muslims living largely on islands in the Mekong delta, have engaged with new markets in a distinct way (Taylor, 2006): men travel widely and across borders as traders serving remote markets, whereas women largely remain at home and look after small agricultural plots. Captured fish are consumed, but rarely sold. The Cham were also largely excluded from the caged catfish aquaculture industry, which emerged in the mid 1990s and was dominated by wealthier ethnic Kinh. This was explained as due to a lack of financial capital, technical expertise and relationships of the resident women with potential investors and buyers. On the other hand roles are sometimes reversed. In Cambodian households in Poipet across from Thailand many women are involved in the cross-border trade of fish. Some husbands without full-time work do housework – taking care of children, washing clothes – contrary to norms elsewhere in rural Cambodia (Kusakabe, 2004).

Rural extension programmes that deliver new techniques for culturing and processing fish often presume that 'free' labour is available at home. Aquaculture based in ponds or cages near the home lends itself to integration with other agricultural activities and household chores. The multiplication of roles and responsibilities can mean extra work burden and reinforce constraints on mobility. Women in the Mekong region often accept these extra burdens, in part because of socialization processes that have created strong expectations about their roles and responsibilities, and partly because they are able, despite these constraints, to extract some benefits from such engagement.

Making Decisions

A critical dimension of empowerment of women is their power to make decisions on matters affecting their lives and those of others such as family or community (Kabeer, 2005; Yukongdi, 2005). In some societies there is an explicit recognition of importance of joint decision-making between husbands and wives (e.g. Hamilton et al, 2001) but in many others major decisions are taken by the 'head of the household', which when both members of a couple are present invariably means a male. Economic development opportunities, however, change who is at home and who earns income, which, in turn, can influence patterns of decision-making (Pingali, 1997; Mills, 2001).

Aquaculture

Kusakabe (2003) interviewed women in 11 household farms in northeast Thailand. Of these, two were intensive cage culture operations using pelleted feeds and sex-reversed tilapia. In both cases the husband of her female informants was the main decision-maker. From the comparison (D=>C transition, see Figure 6.1) Kusakabe concluded that the greater material resources available to women in more commercial-oriented, intensified, aquaculture farm did not lead to greater decision-making power because social norms and expectations remained unchanged.

In the upper Ping, women were often involved, and sometimes dominant, in making investment decisions. By comparing 200 fish farming and 200 non-fish farming households we could probe the effects of engagement in aquaculture (E=>C transition, see Figure 6.1), class and other factors on decision-making power. Class is an important factor: women in wealthier and more educated households had more influence on decision-making (Lebel et al, 2010). This suggests that overall the benefits to women of engagement in commercial aquaculture are likely to vary with socio-economic class. It should be noted that one constraint of this study is that it doesn't say anything directly about 'commercialization' itself as those who took up fish cage culture were often already engaged in other commercial agricultural activities.

Consultation within households is common, but men don't see women as important in decision-making on agricultural investments as women see themselves. Women's views about their own capacities varied, those with a dominant role in farm management were more confident and assertive about their capacity to decide about and farm fish. Households in which women had a larger influence over agricultural investment decisions, however, were not significantly more likely to have ever farmed fish in river cages (Lebel et al, 2010). This implies that fish farming on its own had not detectably altered gender relations or empowered women with respect to decision-making within the home.

Loans and contracts are key decision points for aquaculture. In the upper Ping many farmers entered into contract arrangements with larger firms, in particular, Charoen Pokphand. Before entering contract arrangements husband and wife may discuss options, but men largely negotiate and make the agreements. Although both women and men could and did take out loans in this community, the common view of many stakeholders remained that men, as 'head of a household', should be the main players in taking loans and signing contracts (Lebel et al, 2010).

A study of cage culture in the lower Nam Songkhram River in northeastern Thailand found women are involved to a level similar to men in feeding and rearing fish, but even more in the selling. Women and men make decisions jointly about fish farming (Sriputinibondh et al, 2005). In another study in northeastern Thailand, decision-making changes with inclusion of aquaculture ponds into farming systems. Joint decision-making was more common (78 per cent) on integrated than conventional farms where the husband made most decisions (56 per cent) (Setboonsarng, 2002). In this location women very rarely made decisions alone.

In the upper Ping, information constraints did not appear to be a reason for differences in decision-making roles (Lebel et al, 2010). Women and men both argued that gender didn't make much difference as the information (about fish farming) was the same. At the population level men and women had similar levels of education and training in aquaculture and agriculture and obtained information from similar sources. Experience was seen as much more important than gender in explaining success as a fish farmer, a situation unusual in the Mekong region.

Men made most production decisions in a Vietnamese upland community because they were considered more knowledgeable from contacts with other people and often having a higher education (Kibria and Mowla, 2006). Likewise in southeast Cambodia researchers noted a disadvantage for women was that they were unable to attend trainings as they lacked time and were often illiterate (So et al, 1998). Biases in extension services and typical village-level fish farming groups reduce participation of women in training activities in Laos (Murray et al, 1998).

Capture fisheries

Beyond management of immediate household finances the role of women as decision-makers in rural communities engaged in capture fisheries is often fairly limited. Among the larger decisions recognized as important by households engaged in capture fisheries are decisions to purchase new boats, fishing gear or equipment. Here men make most final decisions although women may also be consulted (MAFF and CBNRM, 2008).

In decisions to enter other fishery-related livelihoods that are less capital-intensive – such as home-based processing fish or production of fishing gear – women, in general, have a larger role than those involving large purchases. This is consistent with the idea that such activities are an extension of immediate household management responsibilities.

Reflecting labour patterns, men usually make decisions about where and when to fish whereas women decide on processing and selling fish (MAFF and CBNRM, 2008). In Tonle Sap fisheries women also have an important role in decisions to lend or borrow money through informal channels, in part, as they involve the women's relationships with others.

In more public spheres, for instance decisions related to community development and natural resource management (see section below for details), women are often under-represented. In Tonle Sap higher social status in community meant men were the main decision-makers, especially in more formal settings (IFM and ADB, 2007). Women's representation in governance bodies is often crucial to inclusion of women's priorities as men are often particularly weak at taking into account special interests and needs of women (Gätke, 2008). Moreover, when they are members, women often hold minor positions with no decision-making role, like secretary or treasurer (IFM and ADB, 2007).

The needs for, as well as opportunities and capacities for, decision-making may be influenced by commercialization and shifts to aquaculture. Decision-making, however, has many domains. Expanding influence in one domain (for example, fish farming practices or selling of harvests) may not easily translate into other domains, such as major asset purchases within the household or relations more widely in the community.

Managing Natural Resources

Water and river management initiatives increasingly also reach out to women (Resurreccion, 2006; Resurreccion and Manorom, 2007). Women are widely portrayed as more responsible managers of water and common pool resources (Jackson, 1993; Cornwall et al, 2007; Leach, 2007; Resurreccion and Manorom, 2007). Thus, a popular notion in Thai culture is that women are in some way closer, or akin, to nature and water: rivers are referred to by the prefix *mae*, meaning mother. These portrayals of closeness and naturalness have strategic values to both men and women, but should be seen for what they are: social products (Jackson, 1993; Cornwall et al, 2007).

Aquaculture

Commercial fish cage culture in rivers, because it takes place in public spaces, also raises issues of river and water resources management (Lebel, 2008). Women who farm fish in river cages, by definition, also expand their access to natural resources – river water surface areas suitable for fish farming (Lebel et al, 2010).

Women's activities overall on integrated farms in Buriram were associated more strongly with on-farm recycling and environmental improvements than those of men (Setboonsarng, 2002). This suggests that more support for women, for example, in training or extension, could also promote ecological sustainability.

Commercialization and transitions from capture to culture may exclude some people from management and use of previously common pool resources. In the Nam Dinh study, gender and class excluded people from more profitable activities (Le, 2008b). Privatization of coastal lands, for example, for shrimp or clam cultivation reduced common property areas upon which women-headed households, women and girls depended most for their livelihoods (Le, 2008b). Women's access to mangrove resources became restricted and strongly mediated through men who controlled new land leases and shrimp businesses. This happened against a recent history where war-related absences meant women's positions in the rural community had expanded (Le, 2006).

Stock enhancement of lakes and ponds in Savannakhet Province, Laos, led to institutional changes in a community fishery. After stocking water bodies, the Village Administration decided who had what sort of rights to fish based on being a member of fishing teams or carrying out community work. Women, however, were rarely involved in these types of activities (Garaway, 2006) and so would appear to have lost some rights or at least failed to gain a share in the benefits. The village sold most of the fish so this change was also an example of commercialization.

Capture fisheries

Women are clearly under-represented in community fisheries committees in Cambodia. The perception of men and women in those committees, however, is that those who do belong make an important contribution (Gätke, 2008). Women are perceived as more committed to common goals and helping the poorest groups. They are also perceived as more skilled in communication and conflict management. Common constraints on greater participation of women in committees included: lack of time due to household duties, lack of education and experience and lack of support from husbands.

Efforts to conserve forest resources around the Tonle Sap Lake create additional pressures on poor households already made vulnerable due to difficulties of access and over-exploitation of fisheries, which in turn help drive changes in division of labour within households and livelihoods needed to cope. Women bear more of these burdens than men. In 2005 the flooded forest around communes was established as a protected area controlled by the Department of Environment and the villagers were not allowed access to these areas (Mith, 2008). Since then, villagers have faced trouble finding wood and fuel wood. In response they sought out alternative income sources such as spending longer hours fishing, increasing the size of their fishing gear, migrating out, raising more domestic animals, as well as selling their labour to large-scale fishers. At the same time a ban was introduced on fish culture in the studied village to prevent over-fishing; collection of feed fish to support cage aquaculture of carnivorous species had added extra pressure on fisheries (Mith, 2008).

Engagement in management of shared natural resources that underpin fishery and aquaculture livelihoods requires interactions with others outside the household. The communication context for negotiations and decisions is thus very different from that within the household and the pressures on gender relations distinct. Markets and political arenas are distinct settings in the Mekong region. In the marketplace, women have engaged successfully in many instances; in more political spheres women have been much less involved, and with some important exceptions, there is less clear evidence of empowerment.

Discussion and Conclusions

Gender relations

Aquaculture and inland capture fisheries are often seen as opposites but this appears to be an oversimplification of the range of social–ecological relationships present. Inclusion of aquaculture in livelihoods may be highly complementary to existing capture fisheries or it may be in direct competition for natural resources and labour. Either way, capture fisheries and aquaculture have distinct features that impact on division of labour and responsibilities between women and men. The benefits accrued to women from their engagement in fisheries and aquaculture activities, however, are less predictable as they are influenced by specific cultural, market and other factors (see Table 6.2). Four factors appear to be important to understanding whether and how engagement leads to material benefits or empowerment: novelty, diversity, mobility and proximity.

Novelty, in terms of new knowledge about techniques or market channels, can increase the status of women within households and the community. Technical

Table 6.2 *Impacts of commercialization and capture-to-culture transitions on gender relations*

Gender relations dimension	Impacts of commercialization	Impacts of capture-culture transition
Contributing labour	Longer commodity chain => more types of positions => some that are women-dominated occupations	On site => more opportunities to be part of activity
Capturing benefits	Cash outcomes=>work more visible => stronger claim on receipts	On site => more opportunities to collect receipts if product sold at side of pond Knowledge of new techniques => earn respect of men
Multiplying burdens	Emphasis on income earning => extra burdens 'acceptable'	Daily care => reduced mobility
Making decisions	Loans/titles/account roles => reinforced by larger/formal investment decisions	On site => more opportunities to influence management/harvest decisions
Managing resources	Expansion => resource competition => new management regimes => exclusion of women if pre-existing institutions and norms are reinforced or differential vulnerability	Fixed sites and single operator => more governable

skills associated with aquaculture production or post-harvest processing can be a basis for women's empowerment. Women's competence, once acknowledged, can be a source of bargaining and decision-making power. In northern Thailand, women who run fish farms are as knowledgeable as men, make similar profits and are respected for their skills (Lebel et al, 2010). Commercialization of fishing and culture goes hand in hand with wider market relations. In rural areas of the Mekong region, selling and trading are usually considered 'traditional' area of women's expertise and thus a foundation upon expansion around new products. Income generation by itself, however, is not always a good indicator of empowerment as control of activity and decisions may still lie elsewhere.

Diversity of income sources is one outcome of novelty, but may subsequently be lost through specialization. Inclusion of commercial aquaculture in the portfolio of activities of an individual or household, at least initially, typically results in diversification of income sources. As a new income stream there is an opportunity for women to capture benefits, especially from their own labour contributions. Commercialization opportunities match current aspirations and needs of many rural women in the Mekong region, for example, for their children to be healthy and well educated. Specialization at individual and household level may affect gender relations in ways distinct from initial diversification into new activities. If norms have not changed women may again be at a disadvantage. In many rural households, non-farm incomes have increased; along with changes in family structure, education levels and labour these factors increasingly determine how culture and capture fisheries activities fit into livelihood portfolios.

Mobility norms for men and women are important for aquaculture and capture fisheries development. Sullivan's study (2006) showed that differences in mobility between men and women may be constraining rates of learning with respect to fish farming. Women take care of fish but do not interact with agents, other experts or growers in other locations any where near as much as men who are more mobile. If women are not allowed to move around but are largely responsible for farming fish then associations and other networks that could support collective action are less likely to form or be strong. This point was also made by Kusakabe (2004) with respect to small fish traders along the Thai–Cambodian border. Mobility is an even bigger issue in capture fisheries. Women in many cultures do not feel safe travelling beyond their communities overnight without male kin – a requirement for fisheries on larger water bodies that require absences for days at a time. Many inland fisheries, on the other hand, are near homes.

Closely related to mobility, therefore, is the issue of proximity. Fish farming requires having someone around the pond or cage several times a day; the timing requirements for fishing depend on gear and are more flexible. Fishing activities can take place a fair distance from the home and involve overnight stays where such practices in aquaculture are rarer. From a governance perspective there are some differences between capture fisheries and aquaculture (Chuenpagdee et al, 2008). Farmers with fish in ponds are typically easier to govern than fishers following mobile stock in the wild. On the other hand small inland fisheries may also take place in relatively fixed locations.

In some situations the intensification of capture fisheries or aquaculture for commercial purposes may have relatively more adverse impacts on women through differential impacts of environmental degradation. Women, for instance, often have major roles in collecting and using other wetland resources apart from target species fished and cultured. Moreover, privatization of formally common pool resources to make shrimp or fish ponds may deny access to poor women and girls for whom access is most critical.

Commercial aquaculture on the other hand may contribute to food security by improving access to low cost food. Kusakabe (2004) noted in her study along the Thai–Cambodian border that cultured fish was exported from Thailand to Cambodia and became increasingly popular with poorer households because it was cheaper than captured fish from Tonle Sap.

While the expansion of aquaculture and its commercialization may lead to changes in gender relationships – including empowerment of women – the converse may also be true. Some of the differences in how aquaculture or fisheries has unfolded in different locations or within particular social classes may reflect inflexible gender relations. Across the Mekong region, religious and family institutions reinforce subordinate roles for women. Empowerment and disempowerment of women are often both plausible outcomes of shifts from one form of making fish to another, whereas transforming pre-existing gender relations is often a more subtle and non-linear process.

Research opportunities

Our review was based on interpretation of published documents. Studies varied in methods, scope and details of reporting. The evidence-base for this review had some important limitations that in turn may also be seen as research opportunities.

First, most studies were based on observations at one time comparing households with different livelihood mixes rather than longer-term studies of individual households as they changed. The frequent emphasis on households as a unit of analysis in aquaculture and fisheries studies has meant that gender relations and women's interests are overlooked (IFM and ADB, 2007). A cross-sectional design also means that inferences about effects of transitions are indirect and easily confounded by other factors. Longitudinal studies would improve understanding of livelihood dynamics, for instance, substitution processes as particular activities shift from being supplementary to primary and vice versa. Historical depth is also important to understanding gender as a dynamic process through which particular activities, roles and responsibilities become seen as masculine or feminine.

Second, most studies in the Mekong region have focused on farmers and fishers. A handful of studies have looked at processing, marketing (Kusakabe, 2004; Kusakabe et al, 2006) or distribution and consumption. Inequalities between men and women occur at different levels and among links in the production-consumption system (Tindall and Holvoet, 2008). The various studies are not yet easily integrated with each other for a more complete gender analysis of a production-consumption system. How fishing and aquaculture are incorporated into individual and household livelihood strategies needs further exploration, in particular how women's and men's interests are served.

Third, few studies have gone much beyond describing division of labour and the consequences this has for overall work burdens. More studies of other aspects like negotiation and decision-making in households and communities is needed. More work is also needed on the strategies through which women in different circumstances expand access to resources and bargaining power or otherwise exercise agency (Agarwal, 1997; Kabeer, 2005; Cornwall et al, 2007). In Mekong region countries this will require taking into consideration how norms that are embedded in religious and family institutions differ and change.

Fourth, most studies on gender have not dealt adequately with intersections of class, ethnicity and other sources of social differentiation. Most research on gender in Vietnam, for instance, represents women as an undifferentiated group (Scott and Chuyen, 2007). On the other hand a large body of otherwise informative research on aquaculture economics largely ignores gender and other distributional issues within households even as levels of household wealth are used to assess development opportunities (e.g. Sheriff et al, 2008).

Finally, more research is also needed on the dynamics and stickiness of social institutions in which fisheries and aquaculture development take place, recognizing that sometimes they may be intransigent, and other times, responsive. Gender as process is affected by many factors. The relationships between gender, different forms of production, management of natural resources and the environment are contingent and dynamic (Nightingale, 2006). Attention to these processes should lead to a better understanding of the conditions under which gender relations can be improved and women empowered.

Policy implications

Most policy on aquaculture and fisheries management and development has not taken gender issues into account (Siason et al, 2002; Choo et al, 2008). In Cambodia, for example, women have historically been neglected in fisheries policies and programmes (IFM and ADB, 2007). The underlying assumption has been that 'women are physically weak and not suitable for fishing, and that technological innovations are not for women.' Capacity building for women is important to build confidence of women to engage in fisheries management activities (MAFF and CBNRM, 2008); but training and gender awareness activities also need to target men.

Once it has been recognized that women are making a large contribution to aquaculture or capture fisheries there is a tendency to rationalize development policy in instrumentalist ways (Resurreccion and Manorom, 2007). A recent

IFM and ADB (2007) report says: '…gender is now a key development issue. In a post-conflict country such as Cambodia, where many households are headed by women and a major part of fisheries production is in the hands of women, poverty will not be reduced if projects and programmes do not equitably address their situation.' Private and public programmes often assume women have under-utilized time (Brugere et al, 2001; Sverdrup-Jensen, 2002; Kelkar, 2005). Private firms also target women, for example to work in their factories, because they are more careful in their work, easier to control and can be paid less than their male counterparts (Brugere et al, 2001; Mills, 2001).

At the regional level gender issues are receiving increasing attention. The Technical Advisory Body for Fisheries Management of the Mekong River Commission (MRC-TAB, 2006) made several policy recommendations based on reviews of research in the region (e.g. Siason et al, 2002; Kusakabe, 2005). At the Ministry or Department level this included calls for policies to take into account the role of women and for senior fishery legislators and policy-makers to be trained on gender issues. At the community level the call is to 'encourage the participation of women' and improve their access to information; these recommendations build on work by the Network for Women and Gender in Fisheries Development in the Mekong Basin (Sverdrup-Jensen, 2002).

Organizational cultures, through their impacts on everyday practices, power relationships and the context in which staff work, translate commitments to empowerment in text into diverse practices (Bebbington et al, 2007). Gender mainstreaming policies of Thai, Cambodian and Lao governments in fisheries, for example, 'evaporated' as they trickled down to lower levels (Kusakabe, 2005). The Vietnam Women's Union, active in providing micro-credit, training and technology, reinforced some pre-existing gender norms about housework even as they attempted to empower women in fisheries and agricultural activities (Schuler et al, 2006). Across policies and programmes a shift is needed from building awareness of gender issues, recognition of rights and equal participation, to empowerment.

A few years ago, women made up less than half of the staff in Ministry/Department's of Fisheries in the Lower Mekong Basin countries: Cambodia (14 per cent), Laos (25 per cent), Vietnam (28 per cent) and Thailand (37 per cent) (MRC-TAB, 2006). Although the situation has improved in recent years, especially in Cambodia, the proportions remain low in all countries at more senior policy-making levels. Greater attention needs to be made to participation of women in policy-making both within the administrative bureaucracy and elected governments.

The gender-informed analysis in this review challenges some commonly held assumptions about the aquaculture–capture fisheries divide and consequences of commercialization within the Mekong region. The fisheries–aquaculture divide has significant gender dimensions. There is now a growing body of evidence that pre-existing gender and class inequalities can shape the form and impact of capital and how women's labour is subsequently used. Engagement in commercial aquaculture or fisheries is not automatically an empowerment pathway or a burden-multiplier: women's bargaining power and circumstances differ, pre-existing gender relations shape how development unfolds in different societies and women's strategies vary from place to place and change with time.

Acknowledgements

The support of International Fund for Agricultural Development and Echel Eau through the Challenge Program on Water and Food for Project PN50 is gratefully acknowledged. Special thanks to Eric Baran, Babette Resurreccion, Kanokwan Manorom, Rajesh Daniel and Nathan Badenoch for constructive feedback on earlier versions of this manuscript.

References

Agarwal, B. (1997) '"Bargaining" and gender relations: Within and beyond the household', *Feminist Economics*, vol 3, pp1–51

Baran, E., Jantunen, T. and Kieok, C.C. (2007) *Values of Inland Fisheries in the Mekong River Basin*, World Fish Center, Phnom Penh, Cambodia, p76

Bebbington, A., Lewis, D., Batterbury, S., Olson, E. and Siddiqi, M. (2007) 'Of texts and practices: Empowerment and organisational cultures in World Bank-funded rural development programmes', *Journal of Development Studies*, vol 43, pp597–621

Bene, C. (2005) 'The good, the bad and the ugly: Discourse, policy controversies and the role of science in the politics of shrimp farming development', *Development Policy Review*, vol 23, pp585–614

Bourdier, F. (1998) 'Health, women and environment in a marginal region of northeastern Cambodia', *GeoJournal*, vol 44, pp141–150

Brugere, C., Kusakabe, K., Felsing, M. and Kelkar, G. (2001) *Women in Aquaculture*, Asia Pacific Economic Cooperation Project FWG 03/99, Institute of Aquaculture, University of Stirling and Gender and Development Programme, Asian Institute of Technology, Bangkok, Thailand, p60

Bush, S.R. (2008) 'Contextualising fisheries policy in the Lower Mekong Basin', *Journal of Southeast Asian Studies*, vol 39, pp329–353

Chankaew K. and Team Ae (1996) *Kaan Kamnod Chan Kunnapaap Lumnam Korng Prathet Thai*, [*Watershed Classification in Thailand* (in Thai)], Kasetsart University, Bangkok

Choo, P.S., Nowak, B., Kusakabe, K. and Williams, M. (2008) 'Guest editorial: Gender and fisheries', *Development*, vol 51, pp176–179

Chuenpagdee, R., Kooiman, J. and Pullin, R. (2008) 'Assessing governability in capture fisheries, aquaculture and coastal zones', *The Journal of Transdisciplinary Environmental Studies*, vol 7, pp1–20

Cornwall, A., Harrison, E. and Whitehead, A. (2007) 'Gender myths and feminist fables: The struggle for interpretive power in gender and development', *Development and Change*, vol 38, pp1–20

Coyle, S. and Kwong, J. (2000) 'Women's work and social reproduction in Thailand', *Journal of Contemporary Asia*, vol 30, pp492–506

Curran, S.R., Garip, F., Chung, C.Y. and Tangchonlatip, K. (2005) 'Gendered migrant social capital: Evidence from Thailand', *Social Forces*, vol 84, pp225–255

Dang, N.B. (2008) 'Fishing as livelihoods and illegal activity: A case study of fishing livelihoods and the management of marine capture fisheries in a community in a province in the South of Vietnam', MA Thesis in Rural Development, Department of Urban and Rural Development, Swedish University of Agricultural Sciences, Uppsala, Sweden, p61

Dixon-Woods, M., Bonas, S., Booth, A., Jones, D.R., Miller, T., Sutton, A.J., Shaw, R.L., Smith, J.A. and Young, B. (2006) 'How can systematic reviews incorporate qualitative research? A critical perspective', *Qualitative Research*, vol 6, pp27–44

Dorairajoo, S.D. (2002) 'No fish in the sea: Thai Malay tactics of negotiation in a time of scarcity', PhD Thesis, Department of Social Anthropology, Harvard University, Cambridge, USA, p391

Dowley, A. (2008) 'Living on water, women's thirst in the Tonle Sap Biosphere Reserve', Senior BSc Thesis, Bachelor of Arts degree in Geography, Vassar College, Poughkeepsie, NY, p202

Dudgeon, D. (2005) 'River rehabilitation for conservation of fish biodiversity in Monsoonal Asia', *Ecology and Society*, vol 10, no15, available at www.ecologyandsociety.org/vol10/iss12/art15/ (last accessed 12 November 2010)

Friend, R.M. (2009) 'Fishing for influence: Fisheries science and evidence in water resources development in the Mekong basin', *Water Alternatives*, vol 2, pp167–182

Friend, R., Arthur, R. and Keskinen, M. (2009) 'Songs of the doomed: The continuing neglect of capture fisheries in hydropower development in the Mekong', in F. Molle, T. Foran and M. Käkönen (eds) *Contested Waterscapes in the Mekong Region: Hydropower, Livelihoods and Governance*, Earthscan, London, pp307–332

Galmiche-Tejeda, A. and Townsend, J. (2006) 'Sustainable development and gender hierarchies: Extension for semi-subsistence fish farming in Tabasco, Mexico', *Gender, Technology and Development*, vol 10, pp101–126

Garaway, C. (2006) 'Enhancement and entitlement: The impact of stocking on rural households' command over living aquatic resources: A case study from the Lao PDR', *Human Ecology*, vol 35, pp655–676

Gätke, P. (2008) 'Women's participation in community fisheries committees in Cambodia', *Environment and Development*, Roskilde University, Copenhagen, Denmark, p118

Goetz, A.M. and Gupta, A.M. (1996) 'Who takes the credit? Gender, power, and control over loan use in rural credit programmes in Bangladesh', *World Development*, vol 24, pp45–63

Hamilton, S., de Barrios, L.A. and Trevalan, B. (2001) 'Gender and commercial agriculture in Ecuador and Guatemala', *Culture and Agriculture*, vol 23, pp1–12

He, Z. (2001) 'Gender roles in pond-fish farming in Yunnan, China', in K. Kusakabe and G. Kelkar (eds) *Gender Concerns in Aquaculture in Southeast Asia. Gender Studies Monograph 12*, Asian Institute of Technology, Bangkok, pp43–56

Hortle, K. and Bush, S.R. (2003) 'Consumption in the Lower Mekong Basin as a measure of fish yield', in T. Clayton (ed.) 'New approaches for the improvement of inland capture fishery statistics in the Mekong Basin' (ad hoc expert consultation held in Udon Thani, Thailand, 2–5 September 2002), FAO RAP Publication 2003/01, Food and Agricultural Organization of the United Nations, Bangkok, Thailand, pp76–82

IFM and ADB (2007) 'Promoting gender equality and women's empowerment: GAD activity for enhancing the role of women in inland fisheries in Cambodia', Institute for fisheries Management & Coastal Community Development, TA CAM 614, Final Report, January 2007, Institute for Fisheries Management and Coastal Community Development and the Asian Development Bank, Hirtshals and Manila, p276

Jackson, C. (1993) 'Doing what comes naturally? Women and environment in development', *World Development*, vol 21, pp1947–1963

Jongudomkarn, D. and West, B.J.M. (2004) 'Work life and psychological health: The experiences of Thai women in deprived communities', *Health Care for Women International*, vol 25, pp527–542

Kabeer, N. (2005) 'Gender equality and women's empowerment', *Gender and Development*, vol 13, pp13–24

Kelkar, G. (2001) 'Gender Concerns in Aquaculture: Women's roles and capabilities', in K. Kusakabe and G. Kelkar (eds) *Gender concerns in aquaculture in Southeast Asia. Gender Studies Monograph 12*, Asian Institute of Technology, Bangkok, pp1–10

Kelkar, G. (2005) 'Development effectiveness through gender mainstreaming: Gender equality and poverty reduction in South Asia', *Economic and Political Weekly*, 29 October, pp4690–4699

Kenney, C.T. (2006) 'The power of the purse: Allocative systems and inequality in couple households', *Gender and Society*, vol 20, pp354–381

Keyes, C.F. (1984) 'Mother or mistress but never a monk: Buddhist notions of female gender in rural Thailand', *American Ethnologist*, vol 11, pp223–241

Kibria, M. and Mowla, R. (2006) 'Sustainable aquaculture development: Impacts on the social livelihood of ethnic minorities in northern Vietnam with emphasis on gender', in P.S. Choo, S. Hall and M. Williams (eds) *Global Symposium on Gender and Fisheries*, 7th Asian Fisheries Forum, 1–2 December 2004, World Fish Center, Penang, Malaysia, pp7–14

Kusakabe, K. (2003) 'Women's involvement in small-scale aquaculture in northeast Thailand', *Development in Practice*, vol 13, pp333–345

Kusakabe, K. (2004) 'Women's and men's perceptions of borders and states: The case of fish trade on the Thai–Cambodian border', *Journal of GMS Development Studies*, vol 1, pp45–59

Kusakabe, K. (2005) 'Gender mainstreaming in government offices in Thailand, Cambodia and Laos: Perspectives from below', *Gender and Development*, vol 13, pp46–56

Kusakabe K., Sereyvath, P., Suntornratana, U. and Sriputinibondh, N. (2006) 'Women in fish border trade: The case of fish trade between Cambodia and Thailand', in P.S. Choo, S.J. Hall and M.J. Williams (eds) *Global Symposium on Gender and Fisheries*, 7th Asian Fisheries Forum, 1–2 December 2004, World Fish Center, Penang, Malaysia, pp91–102

Le, T.V.H. (2006) 'Gender, Doi Moi and mangrove management in northern Vietnam', *Gender, Technology and Development*, vol 10, pp37–59

Le, T.V.H. (2008a) 'Economic reforms and mangrove forests in central Vietnam', *Society and Natural Resources*, vol 21, pp106–119

Le, T.V.H. (2008b) 'Gender, Doi Moi and coastal resource management in the Red River Delta, Vietnam', in B. Resurreccion and R. Elmhirst (eds) *Gender and natural resource Management: Livelihoods, Mobility and Interventions*, Earthscan, London, pp23–42

Leach, M. (2007) 'Earth mother myths and other ecofeminist fables: How a strategic notion rose and fell', *Development and Change*, vol 38, pp67–85

Lebel, P. (2008) 'Managing for sustainability: The livelihood opportunities, social implications and ecological risks associated with fish cage aquaculture in the Ping River, northern Thailand' [in Thai], MSc (Fisheries Technology) Thesis, Maejo University, Chiang Mai, Thailand, p195

Lebel, L., Tri, N.H., Saengnoree, A., Pasong, S., Buatama, U. and Thoa, L.K. (2002) 'Industrial transformation and shrimp aquaculture in Thailand and Vietnam: Pathways to ecological, social and economic sustainability?', *Ambio*, vol 31, pp311–323

Lebel, P., Leudpasuk, S., Lebel, L. and Chaibu, P. (2007) 'Fish cage culture in the Upper Ping River' [in Thai], *Journal of Fisheries Technology*, vol 1, pp160–170

Lebel, P., Chaibu, P., Jaichaichom, B. and Lebel, L. (2008) 'Gender and the culture of tilapia in the Upper Ping River in Chiang Mai and Lamphun Provinces' [in Thai], *Journal of Fisheries Technology*, vol 2, pp168–178

Lebel, P., Chaibu, P. and Lebel, L. (2010) 'Women farm fish: Gender and commercial fish cage culture on the Upper Ping River, northern Thailand', *Gender, Technology and Development*, vol 2, p13

Li, T.M. (1998) 'Working separately but eating together: Personhood, property, and power in conjugal relations', *American Ethnologist*, vol 25, pp675–694

MAFF and CBNRM (2008) 'Gender implications in CBNRM: The roles, needs and aspirations of women in community fisheries. Six case studies in Cambodia', Fisheries Administration of Ministry of Agriculture, Forestry and Fisheries and Community Based Natural resource Management Learning Institute Phnom Penh, p124

Mills, M.B. (1997) 'Contesting the margins of modernity: Women, migration and consumption in Thailand', *American Ethnologist*, vol 24, pp37–61

Mills, M.B. (2001) *Thai Women in the Global Labor Force: Consuming Desires, Contested Selves*, Rutgers University Press, New Brunswick, Canada, p218

Mith, S. (2008) 'Changes in livelihood strategies and gender roles in a fishery-dependent floating community in Tonle Sap Lake, Cambodia', MA thesis in Sustainable Development, Chiang Mai University, Chiang Mai, Thailand, p202

MRC-TAB (2006) 'Gender and fisheries in the Lower Mekong Basin', Mekong Fisheries Management Recommendation No 4, The Technical Advisory Body for Fisheries Management (TAB) of the Mekong River Commission, Vientiane, Lao PDR, p7

Murray, U., Sayasane, K. and Funge-Smith, S.J. (1998) 'Gender and aquaculture in Lao PDR: A synthesis of a socio-economic and gender analysis of the UNDP/FAO Aquaculture Development Project LAO/97/007', Provincial Aquaculture Development Project (LAO/97/007), STS Field Document, FAORAP, Bangkok

Nandeesha, M., Nam, S., Vibol, O., Viseth, H. and Hanglomomg, H. (1996) 'Farmers feed fish and fish feed the farmers', *ILEIA Newsletter*, vol 12, p17

Nhung, P.T.H. (2008) 'Fish raising in cages in the Tam Giang Lagoon: The contribution of aquaculture to livelihood security in two villages', MA Thesis in Rural Development,

Department of Urban and Rural Development, Swedish University of Agricultural Sciences, Uppsala, Sweden, p59

Nightingale, A. (2006) 'The nature of gender: Work, gender, and environment', *Environment and Planning D: Society and Space*, vol 24, pp165–185

Noraseng, P., Hirsch, P., Manotham, S. and Tubtim, K. (1999) 'A report on household level fisheries in four villages of Sanasomboun District, Champassak Province, Lao PDR', Department of Livestock-Fisheries, Vientiane, Lao PDR

Phillips, M.J. (2002) 'Fresh water aquaculture in the Lower Mekong Basin', MRC Technical Paper No 7, Mekong River Commission, Phnom Penh, Cambodia, p62

Pingali, P.L. (1997) 'From subsistence to commercial production systems: The transformation of Asian agriculture', *American Journal of Agricultural Economics*, vol 79, pp628–634

Power, M. (2004) 'Social provisioning as a starting point for feminist economics', *Feminist Economics*, vol 10, pp3–19

Resurreccion, B. (2006) 'Rules, roles and rights: Gender, participation and community fisheries management in Cambodia's Tonle Sap region', *Water Resources Development*, vol 22, pp433–447

Resurreccion, B. (2008) 'Gender, legitimacy and patronage-driven participation: Fisheries management in the Tonle Sap Great Lake, Cambodia', in B. Resurreccion, and R. Elmhirst (eds) *Gender and Natural Resource Management: Livelihoods, Mobility and Interventions*, Earthscan, London, pp151–173

Resurreccion, B. and Manorom, K. (2007) 'Gender myths in water governance: A survey of program discourses', in L. Lebel, J. Dore, R. Daniel and Y.S. Koma (eds) *Democratizing Water Governance in the Mekong Region*, Mekong Press, Chiang Mai, Thailand, pp177–196

Schuler, S., Anh, H.T., Ha, V.S., Minh, T.H., Mai, B.T.T. and Thien, P.V. (2006) 'Constructions of gender in Vietnam: In pursuit of the three criteria', *Culture, Health & Sexuality*, vol 8, pp383–394

Scott, S. and Chuyen, T.T.K. (2007) 'Gender research in Vietnam: Traditional approaches and emerging trajectories', *Women's Studies International Forum*, vol 30, pp243–253

Setboonsarng, S. (2002) 'Gender division of labour in integrated agriculture/aquaculture of northeast Thailand', in P. Edwards, D.C. Little and H. Demaine (eds) *Rural Aquaculture*, CABI Publishing, Oxford, pp253–274

Setboonsarng, S. and Edwards, P. (1998) 'An assessment of alternative strategies for the integration of pond aquaculture into the small-scale farming system of north-east Thailand', *Agriculture Economics and Management*, vol 2, pp151–162

Sharma, C. (2004) 'The impact of fisheries development and globalization processes on women of fishing communities in the Asian region', *SPC Women in Fisheries Information Bulletin*, vol 14, pp27–29

Sheriff, N., Little, D.C. and Tantikamton, K. (2008) 'Aquaculture and the poor – is the culture of high-value fish a viable livelihood option for the poor?', *Marine Policy*, vol 32, pp1094–1102

Siason, I.M., Tech, E., Matics, K., Choo, P.S., Sharif, M., Herwati, E., Susilowati, T., Miki, N., Shelly, A., Rajabharshi, K., Ranjit, R., Siriwardena, P., Nandeesha, M. and Sunderarajan, M. (2002) 'Women in fisheries in Asia', in N.H. Chao-Liao, K. Matics, M.C. Nandeesha, M. Shariff, I. Siason, E. Tech, and M.J. Williams (eds) *Global*

Symposium on Women in Fisheries, 6th Asian Fisheries Forum, 29 November 2001, Kaoshiung, Taiwan, World Fish Center, Penang, Malaysia, pp21–48

So, N., Ouk, V., Hav, V. and Nandeesha, M. (1998) 'Women in small scale aquaculture development in southeastern Cambodia', *Aquaculture Asia*, vol 3, pp20–22

Sriputinibondh, N., Khumsri, M. and Hartmann, W. (2005) 'Gender in fisheries management in the Lower Songkhram River Basin, in the northeast of Thailand', 7th Technical Symposium on Mekong Fisheries, Ubon Ratchathani, Thailand, pp111–120

Sullivan, L. (2006) 'The impacts of aquaculture development in relation to gender in northeastern Thailand', in P.S. Choo, S. Hall and M. Williams (eds) *Women in Fisheries*, Global Symposium on Gender and Fisheries: 7th Asian Fisheries Forum, 1–2 December 2004, World Fish Center, Penang, pp29–42

Suntornratana, U. and Visser, T. (2003) 'Women as a source of information on inland fisheries', in FAO and MRC *New Approaches for the Improvement of Inland Capture Fishery Statistics in the Mekong Basin* (ad hoc expert consultation. Udon Thani, Thailand, 2–5 September 2002), RAP Publication 2003/01, Food and Agricultural Organization of the United Nations, Bangkok and the Mekong River Commission, pp45–50

Suong, N.T.T. (2006) 'Income generation activities of traditional fishing groups: Case studies in Ha Cong and Dinh Cu Villages, in Tam Giang Lagoon, Vietnam', MA Thesis in Rural Development, Department of Urban and Rural Development, Swedish University of Agricultural Sciences, Uppsala, Sweden, p70

SDF and FSF (2005) 'A study on the impact of processing industries on fishing communities in Thailand: The changing role of women in fishing communities, Sustainable Development Foundation and the Fedaration of Southern Fisherfolk', in J. Taguiwalo (ed.) *Intensifying Working Women's Burdens: The Impact of Globalization on Women Labor in Asia*, Asia Pacific Research Network, Manila, pp17–68

Sverdrup-Jensen, S. (2002) 'Fisheries in the Lower Mekong Basin: Status and perspectives', MRC Technical Paper No 6, Mekong River Commission, Phnom Penh, Cambodia, p103

Taylor, P. (2006) 'Economy in motion: Cham Muslim traders in the Mekong Delta', *The Asia Pacific Journal of Anthropology*, vol 7, pp237–250

Taylor, P. (2007) 'Poor policies, wealthy peasants: Alternative trajectories of rural development in Vietnam', *Journal of Vietnamese Studies*, vol 2, pp3–56

Tindall, C. and Holvoet, K. (2008) 'From the lake to the plate: Assessing gender vulnerabilities throughout the fisheries chain', *Development*, vol 51, pp205–211

Tuan, T.H., Xuan, M.V., Nam, D. and Navrud, S. (2009) 'Valuing direct use values of wetlands: A case study of Tam Giang-Cau Hai lagoon wetland in Vietnam', *Ocean and Coastal Management*, vol 52, pp102–112

van Zalinge, N., Nao T. and Sam, N. (2001) 'Status of the Cambodian inland capture fisheries with special reference to the Tonle Sap Great Lake', Inland Fisheries Research & Development Institute of Cambodia, Phnom Penh, Cambodia, p8

Williams, S., Hochet-Kibongui, A. and Nauen, C. (2005) 'Gender, fisheries and aquaculture: Social capital and knowledge for the transition towards sustainable use of aquatic ecosystems', APC-EU Fisheries Research Report 16, European Commission, Brussels, Belgium, p28

Wilson, F. (1985) 'Women and agricultural change in Latin America: Some concepts guiding research', *World Development*, vol 13, pp1017–1035

Yu, X. (2001) 'Indigenous women's knowledge of sustainable fishery', in K. Kusakabe and G. Kelkar (eds) *Gender Concerns in Aquaculture in Southeast Asia. Gender Studies Monograph 12*, Asian Institute of Technology, Bangkok, pp27–38

Yukongdi, V. (2005) 'Women in management in Thailand: Advancement and prospects', *Asia Pacific Business Review*, vol 11, pp267–281

Part III

Competing Demands and Protecting the Rights of the Marginalized

7

Fisheries, Nutrition and Regional Development Pathways: Reasserting Food Rights

Robert Arthur, Richard Friend and Mark Dubois

INTRODUCTION

The Mekong region faces complex and difficult choices about the management of water resources, representing as they do choices about alternative development futures. However, these choices are reduced to, and framed as, a relatively straightforward need to make trade-offs in order to meet development objectives, justified by recourse to a loose rhetoric of poverty reduction (e.g. Friend et al, 2009). In development debates these trade-offs are presented in different ways, but often as between economic progress and conservation and stagnation. At its most extreme, the trade-off is presented starkly as a choice between energy generation, in particular hydropower development, and capture fisheries production.

When framed as such a trade-off, the debate takes on a new, and perhaps unexpected, dimension. Capture fisheries represent a significant and important food source within the region and food is so fundamental to human well-being and development that it is hard to envisage a situation in which it could be traded off or put at risk for some other developmental good. Harder still against a backdrop of world food forecasts that are consistently stressing a global food productivity and food production challenge (e.g. McCalla, 1998; Von Braun, 2009; Godfrey et al, 2010). This is especially the case in the Mekong region where food – or more precisely, food production shortages, inequitable distribution, weak entitlements, malnutrition, stunting and health problems – characterize rural and urban society in Laos and Cambodia (Krahn, 2005; WFP, 2008).

Food is at the heart of international commitments to achieving global development and eradicating poverty. For example, eradicating extreme poverty and hunger is presented as the first Millennium Development Goal (MDG). Moreover, food is increasingly being seen as a fundamental human right, more than just a basic need. In its most basic and readily understood sense, poverty and deprivation are associated with hunger. While definitions of poverty have become broader and often include social and institutional dimensions, food and obtaining a socially adequate diet remain central to them (e.g. Townsend, 1993). But increasingly arguments about food, emerging from primary food producers around the world, are entwined with often competing or conflicting values of development – of how people see and value their natural and social environments and means of production and exchange.

Food is central to the rhetoric of national development policy in Cambodia, Laos and Vietnam and appears within the discourse of Integrated Water Resources Management (IWRM). Given this centrality of food to poverty reduction, the importance of fish as food in the region and the marginalization of fisheries in development debates (see Friend et al, 2009), our interest in this chapter is to look more closely at what is potentially being sacrificed and how fisheries, reframed, could in fact contribute to meeting development commitments in the region. There is much at stake.

The rich aquatic resources of the Mekong that depend on the integrity of the river and its floodplains are central features of regional diets (Claridge et al, 1996; Hortle, 2007; Friend, 2008). Yet these are the resources that are most at risk from hydropower development (e.g. Dugan, 2008; Dugan et al, 2010), and that have been recognized as being so for many decades (Tubb, 1966). Rather than being seen as a source of development and poverty reduction, aquatic resources, the main source of animal protein in rural diets in the lower Mekong region where people face endemic food and nutritional deficits, become framed as a regrettable cost of development that can and should be borne for a greater good (Friend et al, 2009).

There is an established crisis narrative for the Mekong fisheries that underpins regional water resource development (Friend et al, 2009). Despite evidence of the huge productivity across the region and recognition of their current importance and value, the wild capture fisheries are poorly understood, undervalued, marginalized and often presented as facing insurmountable problems and inevitable doom (Loc et al, 2007; Friend et al, 2009). This crisis narrative focuses on the problems that are assumed to be inherent to fisheries by their very nature, while downplaying the current benefits they provide or their development potential. Much of the discourse around fisheries is limited to debates about impacts from development, rather than fisheries' contributions to development (e.g. Barlow et al, 2008; Dugan et al, 2010). Fisheries have come to represent a cost of development that can somehow be mitigated through other, market-governed but largely unspecified or uncertain, food production and exchange systems. In the face of this crisis, options for fisheries development are seen principally in terms of substituting loss in the capture fishery with increased aquaculture production (e.g. Dugan, 2008).

Given the questionable ability of aquaculture to replace lost capture fisheries production (e.g. Friend et al, 2009; Dugan et al, 2010; ICEM, 2010), that such a trade-off can even enter public debate raises serious questions about the very meaning and direction of current water resource development strategies in the Mekong. Not least, it raises thorny questions about who might be the winners and losers of these development directions and choices. While the imagery of food creeps into the rhetoric of hydropower itself, for which there is argued to be a 'thirst' and a 'hunger', it is the food itself, in its most real and intimate sense that is to be sacrificed. In local languages, food imagery is tightly interwoven with local conceptualizations of the environment, natural resources and well-being, largely articulated in terms of sources of sustenance, food and medicine. In the hydropower trade-off debate, it is essential to consider critically whose food and whose food rights are being sacrificed for whose benefit, and what it is that the hungry and potentially hungry are expected to do.

In this chapter we consider these issues by focusing on implications of hydropower development for aquatic resources using the example of Lao PDR. Laos has adopted two key national development priorities: reducing poverty, including eradicating hunger, and establishing the country as the 'battery of Asia' through the development of hydropower. While aquatic resources are central to diets and culture, the wild resources have received little attention in national policy (e.g. MAF, 2006; Bush, 2008). Furthermore, it is argued that the aggregate impact of dam development on the fisheries of the Lower Mekong Basin (LMB) would be less severe if dams were developed higher up in the system, and that the impacts on capture fisheries in Laos might be less than for other countries in the region (Dugan, 2008; Barlow et al, 2008; Baran and Myschowoda, 2009). The potential for a trade-off emerges, within Laos itself, and between Laos and other countries in the region.

FOOD, AQUATIC RESOURCES AND DEVELOPMENT IN LAOS

Food is a critical issue in Laos, and a central element of national development. While levels of malnutrition in the country have been compared to those of Sub-Saharan Africa, the country is endowed with a rich cultural and biological diversity (Krahn, 2005). Food security and nutrition have been endorsed in national policy and there are policies that address or affect national food security situation both directly and indirectly (WFP, 2008).

Among the more direct policies are the National Growth and Poverty Eradication Strategy (NGPES) and the Sixth National Socio Economic Development Plan (2006–2010) (NSEDP). The latter incorporates the commitment to meeting the Millennium Development Goals (MDGs) through an implementation plan aimed at abolishing seasonal hunger (defined as rice scarcity) at the household level by 2010 and to reduce under-five child malnutrition to below 30 per cent by 2015. Policies with less direct effects on food security but that are important given the resources allocated to support them, include the ban on shifting cultivation, land reform, biodiversity and forestry policies, opium eradication and government resettlement initiatives (Gregory et al, 2007; WFP, 2008).

Historically within the national food security policies, the emphasis has been on equating food security with improving rice production and availability (WFP, 2008). Similarly, the Food Security Strategy, under the Ministry of Agriculture and Forestry, also emphasizes increased rice production with a target of producing sufficient rice to meet the necessary calorific intake for the entire population by 2010. Rice is a staple and provides around 80 per cent of calories in rural Lao diets. However, nutritional security is not always the same thing as rice security. Fisheries play a central role in Lao culture and cuisine, providing a range of collective and individual human benefits, e.g. income generation, exchange of fish for food, fish as gifts and as payment in kind for services (Baird, 2001; Noraseng et al, 2001; Garaway et al, 2006; Arthur and Sheriff, 2008), but this range of benefits is less well represented in national development policies. In some cases, sufficient aquatic resources can compensate for rice deficits, such that villagers with a rice deficit may actually have a better nutritional status than those with a rice surplus (Meusch et al, 2003).

While fisheries can generate a range of social and economic benefits, ultimately fish are food and contribute to nutrition and well-being. In Laos this is a vital contribution as aquatic animals including fish are the major source of animal protein for the majority of people (e.g. Kristensen, 2001). There is a wide diversity of aquatic foods available in communities (Meusch et al, 2003; Gregory et al, 2007) with the great majority coming from the wild. Studies like this provide fresh evidence of the role and importance of fisheries (and other wild foods) in diverse livelihoods strategies and challenge the long-standing emphasis on rice production (e.g. WFP, 2008).

Wild fish and aquatic organisms are the most frequently consumed types of meat in Lao diets with annual consumption estimated by to be in the region of 29kg per person (compared to 5kg of beef, 6kg of pork and 5kg of chicken) and supplies 48 per cent of animal protein in the diet (information derived from Hortle (2007); see also Gregory et al (2007) for similar results). Wild fish were found to be consumed on average nearly four times a week and other aquatic animals twice a week (WFP, 2008). While discussions of fish in debates on nutrition often stress their contribution as a source of protein, they are also a critical source of micronutrients and dietary fats. The WFP (2008) study found that almost every second household reported that they had not used oil or lard in the past seven days and only 10 per cent reported using it daily.

Cultured fish was the least frequently eaten meat, consumed on average once a week. While Lao households often keep buffalo and cattle, and these play important roles in agriculture and as savings, their consumption at the household level is limited. Rather, it is the importance of wild fish in the diets of rural Lao that is widely known (e.g. Garaway, 1999, 2006; Hortle, 2007; see Chapter 6, this volume).

Fisheries also play an important role as part of rural household livelihood strategies for dealing with unpredictable declines in food production from year to year. As Meusch et al (2003) note of villagers in Attapeu, harvesting aquatic resources constitutes the main household strategy for dealing with fairly regular years of rice production failures. But, significantly there are no established coping strategies for dealing with periods of aquatic resource production failures.

Achieving food security and reducing malnutrition are important challenges facing the government and people of the country. Wild foods and wild fisheries play a critical role in the nutrition strategies of households and may even underpin the viability of some rural households. This role goes beyond that of a 'safety net' to be a crucial component of complex diversified livelihoods strategies in which each of the components is important and thus makes the overall strategy viable. Fisheries are part of a highly interconnected and dynamic household and community pattern of food production, consumption and exchange. In order to appreciate the significance of fisheries it is essential to see them in this broader livelihood context (e.g. Smith et al, 2005; Arthur and Sheriff, 2008; Friend, 2008).

Rural households utilize a wide range of resources, with households juggling different activities in synchrony with cycles of flood and the seasons. For example, in floodplain areas, a heavy flood might inundate much of the rice crop but at the same time provide a more productive fishery. In such circumstances people fish more intensively, balancing the loss to the rice crop. The patterns of flood are not always predictable but people have created responsive strategies that allow them to take advantage of different environmental conditions. Rural households and rural communities are not simply engaged in utilizing natural resources, but have also developed local strategies for the management of these resources based on detailed knowledge of their own needs and priorities as well as the ever changing environment (Arthur et al, 2005; Arthur and Garaway, 2006; Garaway et al, 2006; Arthur and Sheriff, 2008). Local knowledge and action are the basis for individual, household and collective developmental responses and the capture fisheries of the Mekong region represent a rich array of local management regimes that are highly adaptive to changing social and natural environments (e.g. Arthur and Garaway, 2005; Garaway et al, 2006; Friend et al, 2009). As such, household livelihoods have been argued to be dynamic, diverse and adaptive (Friend, 2008). But it is not only at the household level that we need to look at livelihoods. Just as with household livelihood strategies, the local institutions (i.e. the rules and norms) that result are also diverse, dynamic and adaptive.

Many resources in Laos and the wider region are managed collectively, within and between villages creating rules and norms that are refined according to the particular local context. Examples that we have observed include management at the household, collective household, village and collective village levels (Arthur,

2004; Garaway et al, 2006; Dubois, 2006). Even rice fields are managed as fishery habitats, often providing more nutrition from fish and other aquatic resources than they do from rice (Gregory et al, 1996; Halwart and Gupta, 2004). In a number of cases these regimes include periods (sometimes seasonal or in dry years where there are few alternative places to fish) when the fisheries are open access but this has not diminished their sustainability (Arthur, 2004; Arthur et al, 2005; Garaway et al, 2006). This will be illustrated through two examples from fisheries in Laos: 'community fisheries' in Savannakhet Province and wetland management in the Champhone wetlands of southern Laos.

In Savannakhet province, villages manage small water bodies (typically 1–40 hectares) for collective benefits. Villagers' innovations in collective management of these water bodies creates suites of management options that can be used by the village administration in response to different sets of circumstances and to create different sets of benefits (Arthur and Garaway, 2005; Garaway et al, 2006). In a study of over 40 villages, while it was found that management systems could be classified into three main categories, there was a high degree of local variation as the systems were tailored to the local landscape and social conditions (Arthur, 2004; Garaway et al, 2006). For example, where there were few alternative fishing locations it was seen that the management system would ensure that there was still access by households to fish for household consumption in the managed water bodies through options such as open seasons or gear restrictions. These systems were highly responsive to village circumstances. When income generation to fund village development was prioritized – such as investing in the village clinic – management strategies were oriented to meet these targets whilst also maintaining the viability of the fishery. During other periods, for example rice shortages, management restrictions could be lifted to allow access to fish for all households (Arthur et al, 2005).

The Champone wetlands are a rich natural resource. Local villagers recognize the importance and value of the natural and man-made rivers, streams, reservoirs, ponds and swamps and households and villages will often individually or collectively manage these wetlands. Hong Nong Sim in the district of Sombuli is a seasonally connected natural water body that is considered an important resource and dry season refuge for fish by several of the adjacent villages. These villages collectively agreed not to fish the area. Yet when one of the villages, Dong Naer, wished to finance electrification of the village, the villages collectively agreed to allow for fishing during the dry season for one year only and to use the income from selling the harvested fish to finance electrification. Only certain areas could be fished so that the productivity and role of the water body were not compromised and, once the fish had been harvested, the collective fishing ban would be reinstated. Thus management is not fixed but is allowed to respond to changing needs while still reflecting the longer-term considerations. Similar observations of the flexibility of access to 'conservation areas' at the household level for village-managed wetland areas has been observed in Attapeu province in times of household sickness or bereavement (Dubois, 2006).

In each of these cases the combination of local knowledge, importance of fisheries and the capacity and capabilities of local people resulted in dynamic management strategies adapted and adjusted in response to the changing environment and to changing local priorities and needs. These strategies were not the direct result of project interventions but examples of local innovation with resource management. This is not adaptive management as the concept has been applied in much of the natural resource management literature (see Walters, 1986), but flexible, responsive management to meet specific development priorities, using the resource not just for their food needs but also as a communal food safety net and as a source of collective income. There are many more such cases from across Laos, and indeed the Mekong region, that illustrate how local people can develop locally appropriate management arrangements and strategies at the village and inter-village level. From these examples the case can be made that fisheries contribute to, and are a component of, adaptive livelihoods, clearly contributing to development as well as protecting against vulnerability.

Fisheries and Hydropower

The fisheries of the Mekong have been identified as the main natural and economic resource that will be impacted by the rapid expansion of hydropower development in the Mekong region. Because some 40–70 per cent of the fish catch in the Mekong is estimated to depend on species that migrate long distances along the Mekong mainstream and into its tributaries (Barlow et al, 2008), these catches are likely to be affected by hydropower development as a result of altered flow regimes and the barrier effects of dams (e.g. Hill and Hill, 1994; Barlow et al, 2008; Kummu and Sarkkula, 2008). In general although the impacts of dams on fisheries and food security have been recognized, their significance has been downplayed by some actors, as the following statement by World Bank/ADB (2006, p64) illustrates:

> *Although the impact is certainly not negligible it appears that it could be manageable – by creating new wetlands for fish spawning, improving wetland and fishery management, or, as an extreme measure, accepting a decline in fishery productivity and providing compensation in the form of alternative income sources in agriculture, and/or through cash payment.*

The arguments are essentially that: the impacts will not be as severe as thought and could be manageable; development and resettlement schemes will provide opportunities for aquaculture and agriculture that can mitigate losses in the capture fisheries; and economic development as a result of investment in hydropower will compensate for losses.

The extent to which these impacts are manageable and can be mitigated has begun to be questioned. Possible production losses under current development scenarios could be of the magnitude of up to 880,000 tonnes per year by 2030 (ICEM, 2010); the significance of such impacts on capture fisheries production have been recognized (e.g. Barlow et al, 2008; Dugan et al, 2010) together with the need for effective mitigation (Bird and Phonekeo, 2008; Dugan et al, 2010). However, the effectiveness of possible mitigation measures has been questioned (e.g. Friend et al, 2009) and, as Claridge et al (1996) make clear, the extent of the challenge that needs to be addressed, even just for Laos, is substantial:

> *In most rural parts of the lowland plains, as well as in much of the uplands, fish and other aquatic animals provide between seventy and ninety percent of the animal protein in people's diet. For many of these people, not yet or barely in the cash economy, there is no affordable substitute source of protein.* (ICEM, 2010, p12)

As we highlighted earlier, it is not just the magnitude of the likely impacts but also who the winners and losers might be that is important. In this respect the recent Strategic Environmental Assessment of hydropower development on the Mekong mainstream (ICEM, 2010) identifies in relation to the changes that might be brought about by mainstream dams: 'Mainstream hydropower generation projects would contribute to growing inequality in the LMB countries. Benefits of hydropower would accrue to end consumers, developers, financiers and host governments, whereas most cost borne by poor and vulnerably riparian communities' [sic]. The SEA authors go on to identify fisheries as an important component of these costs: 'In the short to medium term poverty would be made worse by mainstream projects, especially among poor in rural and urban riparian areas. Fishers are over represented in poor and vulnerable LMB communities which would be affected by fisheries losses' (ICEM, 2010, p133).

Considering also the effects of the loss of fisheries production on those negatively impacted, Dugan et al (2010) conclude that there is a need to identify innovations that can assist the poor to better cope with changes arising from the impacts of this dam development: 'In the Mekong investments to identify, develop and apply such innovations are now required urgently. These will need to tailor the planning, design and operation of dams to sustain river fisheries and other ecosystem services' (Dugan et al, 2010, p346).

However, they go on to highlight the risks inherent in this strategy: 'This has so far proved elusive in all other major rivers with hydropower developments similar to those proposed for the Mekong, and doing so in the Mekong presents a formidable challenge' (Dugan et al, 2010, p346).

Given this situation, the key issue for the Mekong in the context of hydropower development is quite simply whether losses in fisheries production can be replaced in ways that meet the needs and interests of the diverse peoples who currently depend on them. Similar concerns have been raised in relation to wider food production systems, with Pretty (2000, p3) for example arguing:

> *It is also clear what is important is who produces the food, who has access to the technology and knowledge to produce it, and who has the purchasing power to acquire it. The conventional wisdom is that, in order to double food supply, we need to redouble efforts to modernise agriculture. After all, it has been successful in the past. But there are doubts about the capacity of such systems to produce the food where the poor and hungry people live. They need low-cost and readily available technologies and practices to increase food production.*

Meeting the challenge of replacing losses in fisheries production is not simply a matter of looking at volumes of production. For example, it is often assumed that aquaculture production could take the place of fisheries, yet where there has been growth in aquaculture it has been to produce for specific markets rather than local needs (e.g. Arthur and Sheriff, 2008). Whether it can be achieved to the extent that substitutes and alternatives will be produced on an affordable and appropriate, scale and available to the extent that wild fish already are is highly questionable. Even if future innovation can meet these requirements there are concerns whether any such substitutes would fit in the same way with food behaviours, the time economy of households and gender and wider community roles in food production and distribution. In addition to quantities of feed and water (De Fraiture et al, 2007; Starr, 2008), production of alternatives are likely to require enormous investments in the kinds of animal production systems and support infrastructure, particularly extension, veterinary services that are already extremely weak in Laos.

Developing specific mitigation measures might be less important if national economic development rather than resource management can provide an effective means to reduce poverty. National economic development measured according to gross domestic product (GDP) is presented as providing for other investments in the national economy that can address poverty (including health and nutrition). While these arguments may appear plausible, when we consider nutrition, the evidence from Laos thus far remains inconclusive. In recent years, the economy of Laos has enjoyed favourable annual growth rates of 7–8 per cent, based largely on agricultural production but also from hydropower and mining activities. This has been a period of unprecedented economic growth and also a period in which concerns have increased about the long-term sustainability of the natural resource base (e.g. Shaw et al, 2007). Examining the statistics for nutritional status over the past 10 years, the World Food Programme indicated that there have been some improvements. From 2000 to the most recent survey in 2006 the percentage of underweight children under the age of five has been reduced from 40 per cent to 37.9 per cent and wasting has been reduced from 15.4 per cent to 7.4 per cent. The figures for stunting however, the most persistent form of malnutrition, was only reduced from 47.3 per cent to 41.2 per cent and is coupled with a high incidence of micronutrient deficiencies. This is a long way from the policy objective of 30 per cent by 2015 and is a relatively small reduction over a period of such strong economic growth (WFP, 2008).

Furthermore, the potential of markets to deliver food to the needy in times of shortage or crisis is not clear-cut. In some cases famines have occurred where there are functioning markets (Mackintosh, 1990) and even while food is being exported (e.g. Swift, 1993). Recent analyses of the effects of the recent food crisis suggest that increased dependency on markets for primary producers can also result in increased vulnerability for poorer households in times of crisis: 'Because developing countries are more integrated within world markets through trade, investment flows, and remittances than in the past, the latest food and financial crises have stronger effects on those countries than during previous crises' (Von Grebmer et al, 2009, p17).

Given the enormity of the risk in terms of food production and the potential that it is the poorer and more vulnerable that could be most affected, we suggest that we need a rights and risks approach to development that explicitly considers food rights and food risks. Such an approach should consider food producers rights first. Rather than assuming that diversified livelihoods strategies and food production systems can be substituted to meet future food needs through future innovation, if the rights of those most at risk take precedence there should be more onus on the developers to provide evidence that effective mitigation measures can be introduced on the required scale prior to project approval being granted.

Discussion

The section above highlights issues that have far-reaching implications for the challenges of national economic and human resource development. Food is so fundamental that we should not be constrained by looking at impacts and mitigation but embrace a different way of looking at food in terms of rights, rather than simply production and supply. The work of Amartya Sen and others (e.g. Swift, 1993; de Waal, 1997) has been highly influential in this respect. Sen's work has been instrumental in analysing the relationship between food production and exchange and famine. His analytical concern for why some people die and others do not, led to significant shifts from focusing solely on food production failures to examining other failures that deny some people access to food, thus shifting focus from the national levels, to household and individual entitlements.

> *Starvation is the characteristic of some people not having enough food to eat. It is not the characteristic of there being not enough food to eat. While the latter can be a cause of the former, it is but one of many possible causes…Food supply statements say things about a commodity*

*(or a group of commodities) considered on its own. Starvation statements are about the **relationship** of persons to the commodity...[these statements] translate readily into statements of ownership of food by persons.* [emphasis in original] (Sen, 1981, p1)

Famine does not occur simply as a direct result of production failures (or food availability declines) but rather entitlement failures in which some sections of the population lack the assets, or are unable to exchange their assets into needed food (Sen, 1981; 1999). For natural resources-dependent households this is extremely significant for it illustrates that the food situation of individual households is determined by the combination of their resource base, production and exchange and critically mediated by the social and institutional arrangements that allow for access to and benefits from food (Swift, 1993; Leach et al, 1999).

Emerging from this type of analysis have been rights-based approaches to common property resource management and sustainable agriculture. Common to these is an interest in people determining their ownership over food production and exchange, including the resources themselves, and the social and institutional arrangements by which these are converted into tangible benefits. Through this they are able to ensure access to quantity, quality and variety in order to meet nutritional and cultural needs and their own development ambitions. A critical example of this is food sovereignty – an approach that is emerging from natural resource-dependent communities and primary food producers themselves in the face of pressures on their resources base, and their own institutions for management, production and exchange (e.g. Fisher and Ponniah, 2003; Pimbert, 2008; Rosset, 2008).

Food sovereignty has been used by farmers, fishers and pastoralist groups to assert their right to the food resources and the natural and social environments on which they depend, that they have historically managed and to which they argue they have an enduring right.[1] From a perspective of food sovereignty, when food production is determined by external market forces and global free trade, the capacity of smallholders to maintain their self-sufficiency and thus their access to their resource base is undermined (see also Mackintosh, 1990). They argue that while this form of agricultural development can be successful at producing aggregate volumes of food, smallholders struggle to compete with larger-scale agribusiness within their own country, often being forced from their land. Dumping of food by countries with excess production has huge impacts elsewhere not only on food production systems, but also on the entire rural economy and society. A more critical analysis suggests that this is very much the purpose behind these policies – of privatization, enclosure and commoditization of land resources (e.g. George, 1974). Food sovereignty once again places food as a central part of the cultures and communities that are producing it. Food sovereignty is therefore very much about values about people, communities and their environments.

But food sovereignty can represent more than just food because it can provide a basis for households and communities to determine their own development pathways. As Smith and Haddad (1999) argue, food availability is one of four factors associated with poverty reduction, the other three being female education, family health improvement and status improvements for women relative to men. Through food sovereignty, food provides an entry point for wider development as well as, through nutrition and health, underpinning it. This, for us, represents 'good' development.

Food sovereignty has emerged from food producers fighting against arguments similar to those presented in support of hydropower development but food sovereignty has not yet been tested in the Mekong. We believe that there is potential in the approach. Fisheries are being managed and are generating benefits in many parts of the Mekong. People are able to identify their developmental needs and aspirations and are able to innovate and to manage to meet these.

Starting with fishery livelihoods and building on local capabilities can provide a means to address food, poverty and vulnerability needs. But that is not to say that fish and fishing can meet all development challenges. Health and access to health service delivery, along with water and sanitation, continue to be weak in rural Laos creating huge burdens on households, and undermining human development at the national level. There is clearly a food dimension to this. Agriculture productivity is often low, and extension services are often weak and overstretched. Additionally, given the importance of natural resources in rural livelihoods, degradation of the resource base and closing access to vital resources is widely identified as one of the main causes of households becoming poor. These represent some of the big challenges for development in the region. However, given that fisheries are such an important regional resource in terms of the scale of fishing activity, consumption of fish and geographical distribution of fishing activity, understanding the context of rural livelihoods, poverty and vulnerability and the critical role that fisheries play should be the basis for rural development policies and investment. Addressing the challenges of food and development is not only about food production. It is also about thinking in terms of how food can address malnutrition and ensuring variety and nutritional content, food behaviours, including the way in which food is utilized and how this affects nutritional content and food wastage. Again, looking at food from a livelihoods perspective takes what people do and why as a starting point and brings in the importance of people's own capabilities so that food production systems fit with local circumstances and challenges.

Conclusions

Food is so fundamental to individual well-being and notions of economic and social development that it is hard to imagine how the suggestion of it being traded-off could ever enter serious debate. This in itself should be a cause for alarm, bringing issues of rights to the fore. Despite this, it is rare that the concepts of food and food rights enter academic or policy debates. For us this aspect of the story of capture fisheries in the Mekong is equally disturbing.

There are clear issues of social justice. Losses to these fisheries resources will certainly have a disproportionate impact on rural and poorer people. But while much of the debate on fisheries and hydropower has been framed in terms of impacts and mitigation, we now find ourselves in the alarming situation of public acknowledgement of the potential scale of impact, but, at the same time, recognition that there are no existing mitigation strategies that could be put in place. Placing faith in the ability of science to identify some future mitigation, such as improved fish passes or improved aquaculture, and accelerating the search for innovative solutions, represents a leap of blind faith that puts the most basic developmental rights of people to food at risk.

Food is at the very heart of what counts as 'good development'. How we see food and the ways in which it is produced, shared, exchanged and consumed reflect how we see the social world and our environment. The world is facing an unprecedented food challenge (e.g. Godfrey et al, 2010). The already high rates of food deprivation and malnutrition have become all the more alarming after the recent global food crisis. There has been a period of historic achievements in food production but, despite this, the world still faces levels of food deprivation that, when seen in the context of obesity in many parts of the world and segments of society, are hard to stomach. Against this deeply disturbing backdrop, the notion that a food source intimately linked to natural ecological patterns so much a part of poor people's diet is being considered as a trade-off of development says a great deal about the values that underpin such development.

But there is another dimension to this story. The huge potential that capture fisheries have for fuelling development, albeit development based on different values, is being missed. Capture fisheries across the Mekong region are being managed and are providing a wide range of benefits. The potential for building on this is completely overlooked. While it is recognized that fisheries production is high and many people are engaged in catching and/or eating fish, there is a sense that this is somehow a barrier to development. The dependence on fisheries is being presented as the cause of poverty of people in the region (e.g. Bush, 2008). Poor people fish, and it is because people fish that they are poor (see also Béné, 2003). Poor people eat aquatic resources but, despite this, they face malnutrition. As we have argued previously, this perception of the region's fisheries is due to a number of critical assumptions about the fisheries and the use of aggregate figures that obscure the issues or fail to capture adequately the complexity of the region's fisheries and the benefits they generate (see Friend et al, 2009 for more details). Because of this enduring crisis narrative of regional capture fisheries, any positive developmental options are seen as lying outside the fisheries sector.

Examples from Laos illustrate that the viability of food strategies depend on a productive agricultural and natural resource base and modes of exchange, but also, critically, local capabilities and institutions. Together these elements constitute viable food strategies, but when one or other is undermined, food strategies are unsettled with potentially devastating consequences. Recognition of this highlights what could be at risk from development that considers production as something to be traded-off or replaced by as yet unproven (and in some cases, unspecified) alternatives. This creates its own production risks, placing people in new arenas over which they have limited control. Moreover, it denies the potential of existing resources, capabilities and institutions as drivers of development and people's rights to determine their own development pathway.

While fishery-based livelihoods do not currently figure within state-led and regional development planning, it is not to say that they are not viable according to other paradigms and values. The problem is deeper than fisheries. It is an issue of social justice and of what we take as the starting point for development. Given the complex and difficult choices facing the region, and the pressing development challenges, we would like to see a greater emphasis on food rights in regional development planning debates and local livelihoods identified as a starting point for development investment, with a focus on local capabilities, strengthening existing livelihoods strategies and fisheries as a driver of development rather than on alternatives and replacing them.

Note

1 Food sovereignty has a long history, appearing central to the 19th-century campaigns against enclosure of common land in England.

References

Arthur, R.I. (2004) 'Adaptive learning and the management of small water body fisheries: A case study in Lao PDR', Centre for Environmental Technology, T.H. Huxley School for the Environment, Earth Sciences and Engineering and Imperial College of Science, Technology & Medicine, London, p437

Arthur, R.I. and Garaway, C.J. (2005) 'Learning in action: A case from small waterbody fisheries in Lao PDR', in J. Gonsalves, T. Becker, A. Braun, D. Campilan, H. de Chavez, E. Fajber, M. Kapiriri, J. Rivaca-Caminade and R. Vernooy (eds) *Participatory Research and Development for Sustainable Agriculture and Natural Resource Management: A Sourcebook*, International Potato Center-Users' Perspectives With Agricultural Research and Development, Laguna, Philippines and International Development Research Centre, Ottawa, Canada

Arthur, R.I. and Garaway, C.J. (2006) 'Role of researchers in support of fisheries co-management', proceedings of 13th Biennial IIFET Conference, Rebuilding Fisheries in an Uncertain Environment, Portsmouth, UK

Arthur, R.I. and Sheriff, N. (2008) 'Fish and the poor', in R.M. Briones and A.G. Garcia (eds) *Poverty Reduction through Sustainable Fisheries: Emerging Policy and Governance Issues in Southeast Asia*, Institute of Southeast Asian Studies, Singapore, pp15–37

Arthur, R.I., Fisher, E. Mwaipopo, R. Irz, X. and Thirtle, C. (2005) 'Fisheries management science programme: An overview of developmental impact to 2005', Final Technical Report, MRAG Ltd, London

Baird, I.G. (2001) 'Toward sustainable management of Mekong River aquatic resources: The experience in Siphandone wetlands', in G. Daconto (ed.) *Siphandone Wetlands*, CESVI, Bergamo, Italy, pp61–74

Baran, E. and Myschowoda, C. (2009) 'Dams and fisheries in the Mekong Basin', *Aquatic Ecosystem Health and Management*, vol 12, no 3, pp227–234

Barlow, C., Baran, E., Halls, A.S. and Kshatriya, M. (2008) 'How much of the Mekong fish catch is at risk from mainstream dam development?', *Catch and Culture*, vol 14, no 3, pp16–21

Béné, C. (2003) 'When fishery rhymes with poverty: A first step beyond the old paradigm on poverty in small-scale fisheries', *World Development*, vol 31, no 6, pp949–975

Bird, J. and Phonekeo, V. (2008) 'Hydropower development in the context of integrated water resources management in the lower Mekong Basin', paper presented at the Mekong Hydropower Yangtze Symposium, available at www.mrcmekong.org/download/Papers/Bird%20Voradeth-%20Mekong%20Hydropower-Yangtze-Symposium-29oct08.pdf (last accessed 19 November 2010)

Bush, S.R. (2008) 'Contextualising fisheries policy in the Lower Mekong Basin', *Journal of Southeast Asian Studies*, vol 39, no 3, pp329–353

Claridge, G., Sorangkhoun, T. and Baird, I. (1996) *Community Fisheries in the Lao PDR: A Survey of Techniques and Issues*, Natural Environment Consulting Ltd, Benalla, Australia, p67

De Fraiture, C., Wichelns, D., Rockstrom, J. and Kemp-Benedict, K. (2007) 'Looking ahead to 2050: Scenarios of alternative investment approaches', in D. Molden (ed.) *Water for Food Water for Life: A Comprehensive Assessment of Water Management in Agriculture*, Earthscan, London and IWMI, Colombo, pp91–145

De Waal, A. (1997) *Famine Crimes: Politics and the Disaster Relief Industry in Africa*, African Rights and the International African Institute in association with James Currey, Oxford and Indiana University Press, Bloomington and Indianapolis, p238

Dubois, M. (2006) 'Community based fisheries management: A rapid review of five villages in Attapeu', Mekong Wetlands Biodiversity Programme Aquatic Resources Management Information Series, Vientiane, Laos

Dugan, P. (2008) 'Mainstream dams as barriers to fish migration: International learning and implications for the Mekong', *Catch and Culture*, vol 14, no 3, pp9–15

Dugan, P.J., Barlow, C., Agostinho, A.A., Baran, E., Cada, G.F., Chen, D., Cowx, I.G., Ferguson, J.W., Jutagate, T., Mallen-Cooper, M., Marmulla, G., Nestler, J., Petrere, M., Welcomme, R.L. and Winemiller, K.O. (2010) 'Fish migration, dams, and loss of ecosystem services in the Mekong Basin', *Ambio*, vol 39, pp344–348

Fisher, W.F. and Ponniah, T. (eds) (2003) *Another World is Possible*, Zed Books, London and New York

Friend, R.M. (2008) 'Wetlands, food security and nutrition in the Mekong – lessons and practical implications', in M. Ounsted and J. Madgwick (eds) *Healthy Wetlands, Healthy People*, report of the Shaoxing City Symposium, Wetlands International, pp109–121

Friend, R.M., Arthur, R.I. and Keskinen, M. (2009) 'Songs of the doomed: The continuing neglect of capture fisheries in hydropower development in the Mekong', in F. Molle, T.

Foran and M. Käkönen (eds) *Contested Waterscapes in the Mekong Region: Hydropower, Livelihoods and Governance*, Earthscan, London

Garaway, C.J. (1999) 'Small waterbody fisheries and the potential for community-led enhancement: Case studies in Lao PDR', Centre for Environmental Technology, T.H. Huxley School for the Environment, Earth Sciences and Engineering and Imperial College of Science, Technology & Medicine, London, p453

Garaway, C.J. (2006) 'Enhancement and entitlement – the impact of stocking on rural households' command over living aquatic resources: A case study from the Lao PDR', *Human Ecology*, vol 34, pp655–676

Garaway, C.J., Arthur, R.I., Chamsingh, B., Homekingkeo, P., Lorenzen, K., Saengvilaikham, B. and Sidavong, K. (2006) 'A social science perspective on stock enhancement outcomes: Lessons learned from inland fisheries in southern Lao PDR', *Fisheries Research*, vol 80, pp37–45

George, S. (1974) *How the other Half Dies*, Penguin Books, London

Gregory, R., Guttman, H. and Kekputhearith, T. (1996) 'Poor in all but fish: A study of the collection of ricefield foods from three villages in Svay Teap District, Svay Rieng', Working Paper 5, AIT AquaOutreach Project Asian Institute of Technology, Bangkok, Thailand

Gregory, R., Phongphichith, T. and Somboun, T. (2007) 'Upland aquatic resources in Laos PDR: Their use, management and spatial dimensions in Xieng Khouang and Luang Prabang provinces', report for the Swiss Agency for Development and Cooperation, Vientiane, Laos, p68

Godfrey, H.C.J., Beddington, J.R., Crute, I.R., Haddad, L., Lawrence, D., Muir, J., Pretty, J., Robinson, S., Thomas, S.M. and Toulmin, C. (2010) 'Food security: The challenge of feeding 9 billion people', *Science*, vol 327, pp812–818

Halwart, M. and Gupta, V. (eds) (2004) *Culture of Fish in Rice Fields*, FAO and the World Fish Center, p83

Hill, M.T. and Hill, S.A. (1994) 'Fisheries ecology and hydropower in the Lower Mekong river: An evaluation of run-of-the-river projects', Mekong Secretariat, Bangkok, Thailand

Hortle, K.G. (2007) 'Consumption and the yield of fish and other aquatic animals from the Lower Mekong Basin', MRC Technical Paper No 16, Mekong River Commission, Vientiane, Lao PDR, p87

ICEM (2010) 'MRC Strategic Environmental Assessment (SEA) for hydropower on the Mekong mainstream', Hanoi, Vietnam

Krahn, J. (2005) 'Improving livelihoods in the uplands of the Lao PDR', Upland Food Security and Nutritional Diversity Sourcebook, National Agriculture and Forestry Research Institute, National Agricultural and Forestry Extension and National University of Laos, Laos

Kristensen, J. (2001) 'Food security and development in the Lower Mekong River Basin: A challenge for the Mekong River Commission', paper presented at the Asia and Pacific Forum on Poverty: Reforming Policies and Institutions for Poverty Reduction, Asian Development Bank, Manila, 5–9 February 2001

Kummu, M. and Sarkkula, J. (2008) 'Impact of the Mekong river flow alteration on the Tonle Sap flood pulse', *Ambio*, vol 37, pp185–192

Leach, M., Mearns, M. and Scoones, I. (1999) 'Environmental entitlements: Dynamics and institutions in community-based natural resource management', *World Development*, vol 27, no 2, pp225–247

Loc, V.T.T., Sinh, L.X. and Bush, S. (2007) 'Transboundary challenges for fisheries policy in the Mekong Delta: implications for economic growth and food security', in T.T. Be, B.T. Sing and F. Miller (eds) *Challenges to Sustainable Development in the Mekong Delta: Regional and National Policy Issues and Research Needs*, The Sustainable Mekong Research Network (Sumernet), Bangkok, pp99–143

Mackintosh, M. (1990) 'Abstract markets and real needs', in H. Bernstein, B. Crow, M. Mackintosh and C. Martin (eds) *The Food Question: Profits Versus People?* Earthscan, London, pp43–53

MAF (2006) 'The National Strategy for Fisheries from the present to 2020, Action Plan for 2006 to 2010', Ministry of Agriculture and Forestry Vientiane, Lao PDR

McCalla, A.F. (1998) 'Food needs to 2025', in C.K. Eicher and J.M. Staatz (eds) *International Agricultural Development*, 3rd Edition, JHU Press, Baltimore and London, p597

Meusch, E., Yhoung-Aree, J., Friend, R. and Funge-Smith, S. (2003) 'The role and nutritional value of aquatic resources in livelihoods of rural people: A participatory assessment in Attapeu Province, Lao PDR', FAO, Bangkok

Noraseng, P., Manotham, S., Pransopha, K., Hirsch, P. and Tubtim, K. (2001) 'Management of backswamps in Sanasomboun and Phonethong Districts, Champassak Province, Lao PDR', SWIM Report 1, The Small Scale Wetland Indigenous Fisheries Management Project (SWIM), Australian Centre for International Agricultural Research and International Development Research Centre

Pimbert, M. (2008) *Towards Food Sovereignty: Reclaiming Autonomous Food Systems*, International Institute for Environment and Development, London

Pretty, J. (2000) 'Food security through sustainable agriculture', paper presented at Novatis Foundation for Sustainable Development Symposium, 'Nutrition and Development', Basel, 30 November 2000

Rosset, P. (2008) 'Food sovereignty and the contemporary food crisis', *Development*, vol 51, no 4, pp460–463

Sen, A. (1981) *Poverty and Famines: An Essay on Entitlement and Deprivation*, Clarendon Press, Oxford, UK

Sen, A. (1999) *Development as Freedom*, Oxford University Press, Oxford

Shaw, S., Cosbey, A., Baumüller, H., Callander, T. and Sylavong, L. (2007) 'Rapid Trade and Environmental Assessment (RTEA) National Report for Lao PDR', International Institute for Sustainable Development

Smith, L.C and Haddad, L.J. (1999) 'Explaining child malnutrition in developing countries: A cross-country analysis', IFPRI Research Report 111

Smith, L.E.D., Nguyen Khoa, S. and Lorenzen, K. (2005) 'Livelihood functions of inland fisheries: Policy implications in developing countries', *Water Policy*, vol 7, pp359–383

Starr, P. (ed.) (2008) 'Interview: Jeremy Bird, CEO of the MRC Secretariat', *Catch and Culture*, vol 14, no 1, pp4–7

Swift, J. (1993) 'Understanding and preventing famine and famine mortality', *IDS Bulletin*, vol 24, no 4, pp1–16

Townsend, P. (1993) *The International Analysis of Poverty*, Harvester Wheatsheaf, Hemel Hempstead, p291

Tubb, J.A. (1966) 'A consideration of the fisheries problem of the Lower Mekong Basin', *Indian Journal of Power and River Valley Development*, The Mekong Project, vol XVI, pp62–64

Von Braun, J. (2009) 'Responding to the world food crisis: Getting on the right track', IFPRI 2007–2008 Annual Report Essay, Washington, DC

Von Grebmer, K., Nestorova, B., Quisumbing, A., Fertziger, R., Fritschel, H., Pandya-Lorch, R. and Yohannes, Y. (2009) *Global Hunger Index: The Challenge of Hunger: Focus on Financial Crisis and Gender Inequality*, International Food Policy Research Institute, Bonn, Washington DC, Dublin

Walters, C.J. (1986) *Adaptive Management of Renewable Resources*, Macmillan, New York

WFP (2008) 'Lao PDR: Comprehensive Food Security and Vulnerability Analysis (CFSVA)', Strengthening Emergency Needs Assessment Capacity (SENAC) World Food Programme, Lao PDR

World Bank/Asian Development Bank (2006) 'Future Directions for Water Resources Management in the Mekong River Basin', Mekong Water Resources Assistance Strategy, World Bank/Asian Development Bank, June

8

Livelihood and Environment Trade-offs at the Time of *Doi Moi*: Industrial Water Use and Wastewater Management in a Craft Village in Peri-urban Hanoi

Le Thi Van Hue and Edsel E. Sajor

Since 1986, Vietnam's economic reforms or *doi moi* have dramatically improved the country's living conditions. Rapid growth in both industrial and agricultural sectors has contributed to more than half of the country's gross national product (NEA/WB/DANIDA, 2002). Since 1986, living standards have improved gradually both in urban and rural lowlands, although poverty still persists in the northern mountains, north central coast and the Central Highlands. Economically, Vietnam has progressed significantly in recent years. The country has averaged a GDP growth rate of 8 per cent a year along with reducing poverty from nearly 60 per cent of the population in 1993 to 16 per cent in 2006 (VDR, 2008).

While Vietnam's economic reforms have undoubtedly resulted in a major expansion of industrial and agricultural outputs and in an overall reduction of the poverty rate (Dang, 2009), these have also posed serious problems and challenges to the state of the environment. Several scholars (Jamieson et al, 1998) have pointed out that the country's strategy has implied a drive towards optimal utilization of the country's natural and human resources for fast-track economic growth and the subordination of long-term environmental concerns. But ironically, unlike other countries in Southeast Asia, Vietnam entered this period of catch-up industrialization and modernization with a large catalogue of unresolved environmental problems such as degradation of arable lands and air, water and dust pollution. It is thus forced to seek a delicate balance between economic growth and environmental concerns (Di Gregorio et al, 2003; Sinh, 2004; O'Rourke, 2004). Economic growth too often becomes the overriding priority to the exclusion of environmental considerations despite official rhetoric to the contrary (Kelly et al, 2001).

One of the key features of Vietnam's ongoing economic reforms has been private sector development and its enhanced integration with the global economy. It is in this sector that tension between rapid economic development on the one hand and the environmental concerns on the other is being intensely played out. In Vietnam, craft villages are part of the private sector. There are presently 1480 craft villages in Vietnam, of which 67.3 per cent are located in the north, 20.5 per cent in the Central region and 12.2 per cent in the southern region of the country (MONRE, 2008). The majority of craft villages (up to 80 per cent) are household-based artisanal production. Craft villages create employment for 11 million people, which account for 30 per cent of the labour force in the rural and semi-rural areas. Products are exported to nearly 100 countries around the world. Total nationwide revenue generated from exported products from craft villages was US$650 million and US$730 million in 2006 and 2007 respectively (MONRE, 2008).

This chapter examines the tension in the water domain through primary research of industrial water use and wastewater management in a craft village in Vietnam. The private production case examined in this study is a most pervasive mode of privatized industrial production in Vietnam – the household-based artisanal production – that is linked to domestic and international markets. The particular craft production (metal manufacturing), which is the subject of this chapter, is not only a popular form of livelihood in peri-urban and rural areas but incidentally performs an important environmental function of recycling – paradoxically – while itself creating new and heavy environmental and health burdens in the water domain. The analysis of the chapter pays explicit attention to: a) trade-off between development and environmental protection; b) the effects of national economic reforms on control and access to water resources; and c) the impacts of water contamination on the environment and villagers' health.

The structure of this chapter is organized as follows: the first section discusses the broader context of worsened water pollution, privatization reforms and environment in Vietnam. The second section provides an overview of the study site and a discussion of the history of development of the Van Mon commune. The third section describes the metal recycling process, water use and water pollution in the village. The fourth section presents a profile of villagers' craft livelihood and its impact on the environment and health. The concluding section analyses the key issues with regards to the tension between governance, livelihoods and environment problems and explores an alternative approach to better wastewater management for the commune.

THE BROADER CONTEXT OF WORSENED WATER POLLUTION: PRIVATIZATION REFORMS AND ENVIRONMENTAL PROTECTION IN VIETNAM

A keystone policy of economic renovation (*doi moi*), endorsed by the Vietnamese Communist Party at its 6th Party Congress in December 1986, was a shift away from the command economy model toward a market-based system. A major feature of this policy was allowing households, the bulk of which were engaged in rural farming, the right to choose their own crop to plant, and/or craft activity to engage in. This enabled them to sell their products directly in the market, and to appropriate privately the returns and rewards from these economic activities (Luong, 2003; Kerkvliet, 2005). This meant countrywide de-collectivization and the reinstitution of the family and the household as the most important economic unit, especially in the rural areas. This development would, in the years following the *doi moi* policy, become an important feature of the privatization trend in the country.

Thus, in the 1990s privatization in Vietnam was characterized by rapid expansion of household enterprises rather than the growth of the domestic corporate sector (Gainsborough, 2004). In a 10-year period, from 1988 to 1998, handicraft households among the industrial sectors' categories of the country registered a growth of 318,555 to 553,043 household units (Luong, 2003). These enterprises were mostly concentrated on trade, repairs and personal services. Craft production for export was also a major concentration of the private household enterprises. The range of household or micro-enterprise production had several things in common: their competitive advantage, low technical-technological requirements and relatively low-start up costs (Giao and Cuong, 1995).

By 2001, the number of formal private enterprises had significantly increased. foreign direct investments (FDIs) too rose sharply, mainly in the form of joint ventures with state enterprises as local equity partners, pushing the private sector development to a new level. Despite the growth of these formal private companies, however, household enterprises registered in 2007 numbered about 2.7 million comprising a significant 13 per cent of Vietnam's GDP (VietNamNet, 2007). While the registered output contribution of the household enterprises segment of the private sector might be much less than that of state-owned establishments (41 per cent of output share), and FDI enterprises (35 per cent of output share), its major importance lies in terms of human resource and geographic employment absorption. The household enterprises not only employ a large number of people, but also absorb and make more productive otherwise economically stagnant rural and peri-urban areas of the country. Furthermore, its significance is much more when one considers that, compared to the other private or non-state business organization categories, household enterprises have huge intractable, non-registered household entities (Muller, 2005). Further, formal distinction between household enterprises becomes blurred and the size of non-registered households needs to be adjusted upwards when one considers that many farming households are also engaged in craft enterprises. According to the surveys used to assess poverty in the country in 2005, 75 per cent of households are involved in farming, 38 per cent of which run a small business of one sort or another (and a third are not related to farming) (VCG, 2005). While accounting for these activities is certainly difficult, the data from the Vietnam Consultation Group (VCG) survey clearly indicates that entrepreneurial activities of farming households are thriving.

Alongside privatization reforms that aimed to boost industrial production, it is noteworthy that Vietnam has considerably improved its policy framework for environmentally sustainable development. Since the mid 1980s, the Vietnamese government has become more active in the field of environmental protection. Through preparatory research projects and capacity building in regulatory and monitoring systems, creation of nature reserves began to culminate in putting in place a legal framework for basic environmental regulations. The Law on Environmental Protection (LEP)[1] lays out the rights, responsibilities and structural relationships between ministries, political-administrative units, economic units and individuals through which the law will be carried out (Di Gregorio et al, 2003, pp194–195). Concrete progress on the ground, however, has been less satisfactory (VCG, 2005, p122).

Particularly in the water sector, the Government of Vietnam has issued a number of laws and regulations to protect water resources. Instruction 487/TTg dated 30 July 1996 strengthened the state management of water resources. The Law on Water Resources was enacted in 1998, followed by Decree No-179/1999/ND-CP specifying the implementation of this Law. However, there have been serious limitations regarding the implementation of the Laws. The Provincial Departments of Science and Technology (DOSTE) have identified over 3300 polluting enterprises that cause water and air pollution (NEA/WB/DANIDA, 2002). For rural areas, the National Program for Clean Water and Environmental Sanitation in rural areas was approved in 1998. State targets for 2005 include providing 80 per cent of the rural inhabitants with access to clean water and 50 per cent with hygienic latrines. Nevertheless, the majority of the artisanal villages presently do not have facilities to treat their wastewater. Most of the wastewater is directly discharged into fish ponds, canals and rivers without treatment, which in turn adversely affects local people's health and the environment.

Compliance and enforcement of environmental regulatory standards of private industrial companies remain problematic as found in the studies mentioned below on the environmental performance of corporate industries. O'Rourke (2002) discussed how air pollution, in the form of boiler gases, soot and dyes, by private industrial companies operating outside export processing zones, has caused respiratory problems among residents of neighbouring residential communities. Minh (2002) found out that considerable environmental pollution is also committed by private industrial multinational companies operating mostly in industrial zones and export processing zones, where regulations and environmental infrastructures are better than those of private companies located in unplanned and non-industrial or residential areas. However, even in these industrial and export processing zones, as for example in the case of Ho Chi Minh City, the operations of wastewater treatment facilities have not been managed well or consistently. It has been argued that because industrial zone management boards are under pressure to attract new enterprises and keep existing ones by any means possible, they frequently expedite issuing of permits including environmental impact assessments (EIAs). Further, though mandated to comply with environmental regulations, management boards often act as 'screens between DOSTE, and the environmental monitors and individual firms within their sites' (Di Gregorio et al, 2003, p182). Intervention by management boards has often blunted compliance by industrial companies to regulatory standards set. These deficits in wastewater management by corporate industries particularly in Ho Chi Minh City have contributed hugely, together with untreated domestic wastewater, to a critical level of pollution of the Saigon River (Sajor and Thu, 2009).

What has not been covered by recent studies related to privatization and environment in Vietnam is the domain of household-level manufacturing or craft production (for a few exceptions, see DiGregorio et al, 1999; Ha et al, 2008). There are important reasons though why this should be an important focus of study of privatization and environment in Vietnam. First, as mentioned earlier, private household-level craft production continues to constitute a most pervasive and important aspect of privatization and economic reforms, particularly in terms of providing employment in the rural and peri-urban regions. Second, as economic units and actors, households and its members are known to act in a manner that puts into consideration basic livelihood interests and gains first, while subordinating environmental and health costs. Third, the diffused character of this micro-scale manufacturing process creates environmental burdens both to the producer and non-producer households and entire communities. Therefore disentangling residential and manufacturing functions that combine within a household poses a special challenge to spatial planning and community-wide development. Fourth, effectively managing pollution in craft villages, particularly industrial wastewater, implicates the issue of conflict-laden relationships between the rights of households to livelihoods, the rights of communities or settlements to a livable and healthy environment and the rights and duty of the state to safeguard 'public' water resources from degradation.

Man Xa craft village

Craft villages have had a long history and role in Vietnamese national development. They are a typical feature in the social, economic and cultural tradition of Vietnam, particularly in the countryside and, more recently, in the peri-urban areas of the country. They have made a significant contribution to economic development and to changes in the national economic structure, especially in terms of enhancing local incomes and employment in the villages. Hence government policy-makers have put special emphasis on the development of craft villages as part of their employment and rural development strategy (DiGregorio et al, 1999; Ha et al, 2008). Over the past decades, many craft villages have started to recycle waste materials. In Vietnam today, there are three types of recycling craft villages: plastic recycling, metal recycling (foundry villages) and paper recycling villages.

In the Red River Delta region, after the introduction of economic reforms, many villages redeveloped their specialized craft occupations as a means of improving their livelihoods. These villages had traditionally been engaged in artisan, craft and trade activities for decades in addition to their agricultural activities. But the majority of these activities disappeared during the cooperative regime of the pre-*doi moi* era. However, since the 1980s, these occupations, customarily grouped into handicrafts and small craft industries, have grown on a par with other industries. According to government sources, craft villages have recently exhibited growth rates (measured in the value of output) from 8.8 per cent to 9.8 per cent (MONRE, 2008). In the provinces surrounding Hanoi, where a few important village industry clusters dominate, provincial rates of growth have generally been higher. Between 1996 and 1997, there were more than 178 craft villages and commune clusters in Ha Tay, Bac Ninh and Hung Yen provinces (Ba, 1997; Cuong and Nguyen, 1998). In some villages, these activities have become 'traditional craft' with valuable products such as paper, iron/copper products or sculpture in Bac Ninh province, plastic tools in Hung Yen province and pottery in Hanoi City, among others. These non-farming activities generate jobs for local farmers, especially during leisure time and slack periods for farm labour. These industrial villages, as both processors and producers, have provided much of the demand for recycled materials. The virtuous cycle of rising demand and increasing supply has, in effect, meant that nearly all recyclable materials that appear in Hanoi's economy eventually find their way into production (DiGregorio et al, 1999).

The study site

Van Mon commune located in Yen Phong District, Bac Ninh province is about 21km northeast of Hanoi and 7km southwest of Tu Son town. Van Mon is bordered in the north by Yen Phu commune and Cho town; in the south by Huong Mac, Tu Son; in the east by Cho town and Dong Tho commune and in the west by Thuy Lam commune, Dong Anh, Hanoi (see Figure 8.1). The commune is accessible by roads and waterways.

Van Mon has five villages: Quan Do, Quan Dinh, Man Xa, Phu Xa and Tien Thon. It has a total area of 424.84ha, of which 268 ha is agricultural land, 65.1ha is residential land, 91.3ha is special use land and 0.38 ha is unused land. Van Mon

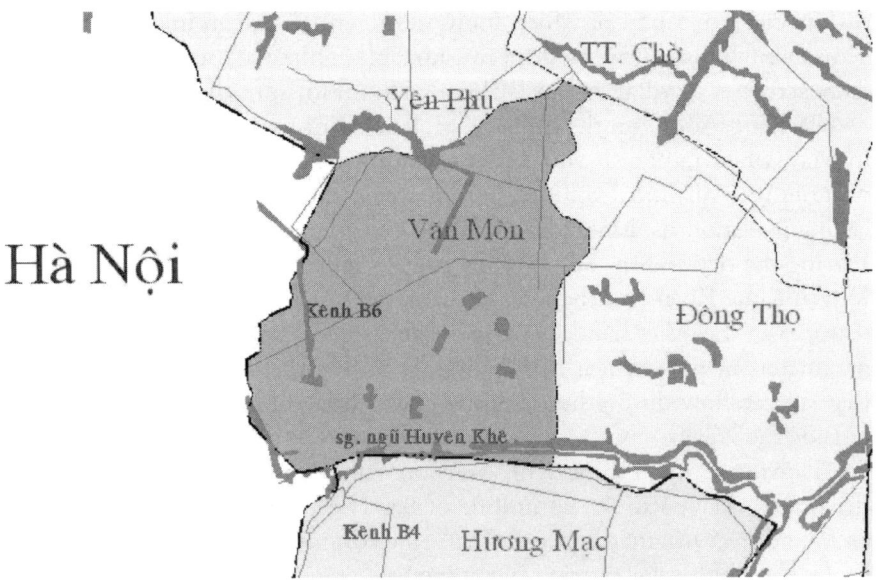

Figure 8.1 *Location of Van Mon commune*

Source: People's Committee of Van Mon Commune, 2006

has one primary and one secondary school, one health clinic and one market that is open every day.

The commune has a population of 9359 people and 1709 households (based on the 2005 census). The annual population growth rate is 1.65 per cent. Of the total population, 3762 or about 40 per cent are of working age. Seventy per cent of adult workers are engaged in artisanal production and rice production and 30 per cent are purely engaged in farming rice. On average, per capita rice production is 167kg of milled rice every six months.

Van Mon has a long history and culture. Elderly individuals within the commune have experienced life under three regimes: the French colonial government, the Japanese occupation and independent Vietnam. During the war, not only did women actively participate in agricultural production, they also joined the army to fight against the US military forces along side the men. During the *doi moi* reforms, a tradition of hard work has been brought into play in the market economy to improve their livelihoods and household economy.

The average rainfall in the area varies between 1240 and 1598mm per year. The rainy season often coincides with the prevalent period of the southwest or southeast wind (between May and October) accompanied by atmospheric turbulences (including tropical convergent strip, typhoon, tropical low pressure), creating long lasting medium and heavy rains (DONRE, 2005). Rainfall during this period makes up about 75–80 per cent of the total annual rainfall. Rainwater creates a surface current, a part of which infiltrates to enrich the groundwater in the area. Thus, rainfall not only provides a considerable volume of water for production and daily activities of villagers but is also a medium for spreading pollution.

The dry season, on the other hand, lasts six or seven months from November to May, when rainfall is very little, accounting between 15 per cent and 20 per cent of the total annual volume. In some years, there are no rains at all for a period lasting 3–4 months. March has the lowest evaporation level in the year of 67mm. During the dry season, oil and iron concentrations in certain points of the Ngu Huyen Khue River and the sewage canals are much higher than during the rainy season. On the other hand, as we will elaborate in a later section, for certain other parameters of pollution it may be higher during the rainy season, an indication that surface flow during heavy rains collects others pollutants from a wider area outside the village.

Flowing through the area of Van Mon with a length of about 2km and joining the Cau River in Van An commune of Bac Ninh province is Ngu Huyen Khue, an inland river originating from Chau Khe commune (Tu Son district). The river provides water for five districts including Yen Lang, Dong Anh, Tu Son, Yen Phong and Tien Du. The Ngu Huyen Khue River is 50–70m wide with a water flow of 60m^3/second. In the rainy season, the water level of the river fluctuates from 3–10m depending on the area. The river's water is used for irrigation purposes. It also receives waste sources of various types from the area.

Development of Van Mon's craft villages

Aluminium melting started under the French era in the neighboring province of Bac Ninh, in Hiep Hoa, Bac Giang, but it was easily adopted by Man Xa village, which has a long history of craftsmanship.[2] From here it spread to the other villages of Van Mon commune. It is said that Mr Hoang Duc started the craft by making pans from the body of an American airplane shot down in the village in 1963. (These pans later came to be named after him.) Then in 1965, aluminium melting was developed and spread to the whole village, and by 1967 an aluminium-melting cooperative became established. With the cooperative's formation, households no longer produced pans on their own. Instead, they were formed into production brigades, which were under the management of the cooperative.

However, the collective model revealed many limitations and mistakes, such as poor management skills of cooperative cadres, poor distribution of goods and reduction in cooperative members' income. Therefore, in the early 1980s with the collapse of the cooperative model, household-based artisanal production increasingly displaced cooperatives (DiGregorio et al, 1999; Luong, 2003).

Years prior to the adoption of the *doi moi*, many villages throughout the country had already been shifting to household-based private economic activities, resulting in less priority being given to collective brigades. In many cases, this had the tacit sanction of local authorities (Kerkvielt, 2005). For example, in 1982 many villages in Van Mon village decided to separate from the cooperative and privately invest in aluminium and metal melting to develop their household economy. In addition, a number of households in the village also conduct lead and zinc melting. In 1995 junk trading was also started and developed in Man Xa. Villagers usually buy junk from other provinces and after sorting it out they sell to households that are engaged in melting or to people elsewhere outside the commune. In the last five years, Man Xa has expanded its markets to China. Villagers began melting aluminium bars for factories, plants and to export instead of making pans for the domestic market. This is partly because more and more households in Vietnam today use electric rice cookers instead. But more importantly, aluminuim bars bring higher profits. These, in fact, significantly contribute to improving their household economy and living conditions after *doi moi*.

Metal recycling, water use and wastewater and other pollution discharges

Presently, Man Xa village is heavily engaged in aluminium and colour metal melting. The main materials used for production of these bars are various types of aluminium scraps. Over 200 households in the village are engaged in this industrial activity. Table 8.1 below shows figures of the aluminium outputs of Man Xa. To produce these outputs, 8000 tons of aluminium scraps and 1200–1500 tons of fuel are used per year.

Aluminium scraps vary in type and size. They can be divided into two main groups: consumer product aluminium (e.g. empty cans, aluminium frames, pots and pans, etc.) and production aluminium scraps (e.g. parts or pieces of machines or equipment made of aluminium alloy).

Both have been used by local households. But the main material group to be used depends on the scale of household production. For medium- and large-scale producer households, consumer product aluminium scraps are commonly used. Although these types of scraps are more expensive, aluminium recovery is relatively greater, at 80–90 per cent, with small amount of slag. For smaller-scale producer households, which have less operational capital, production aluminium scraps, mostly alloys, are their choice, since these type of scrap materials cost less. However, the aluminium recovery rate from this type is lower, about 50–60 per cent, and with a lot of slag.

Table 8.1 *Outputs of aluminium production in Man Xa*

Product	Output per year (ton)
Melted aluminium	400–500
Metallurgic aluminium	4000–5000

Source: DONRE (2005)

Aluminium scraps are brought by scrap merchants in trucks to the village. These are then purchased and sorted out by artisan households. However, merchant buyers also regularly purchase the aluminium bars from the households. They are transported by trucks from the village, where they are purchased by factories in the cities or exported abroad as semi-processed raw materials.

In household-based aluminium melting, the technology is considered extremely inefficient. This is because the households are consuming coal dust and fossil coal, which is 80–100kg/100kg of aluminium products. With an average production of 80kg per household per day, and with 70 per cent of households involved in production, the village utilizes 17,000kg of coal each day.

The aluminium production process in the household follows seven sequential steps (see Figure 8.2):

- Step 1: Washing and preliminary treatment of scraps (this step uses water and produces waste water)
- Step 2: Melting of scraps in primary pots
- Step 3: Pouring of melted aluminium into primary moulds (rough moulding)
- Step 4: Transfer of moulded aluminium to a secondary pot for continuous melting
- Step 5: Pouring of molten aluminium from secondary pot to moulds for forming bars
- Step 6: Cleaning of moulded bars in a basin
- Step 7: Finished bars are prepared for sale and transported to the market.

Water is used and wastewater is generated in washing of scrap materials (Step 1); in filtering slag for further recovery of residual aluminium contents (auxiliary of Step 2); and in cleaning moulded bars in a basin (Step 6), just before they are ready for selling. Solid waste is also produced in the form of dumped slag from melting in the secondary pot (Step 4). Further, a large volume of toxic gases of various types is formed and dispersed to the surrounding environment.

In Man Xa, water used for manufacturing comes from drilled wells and irrigation canals. In the past, a fairly large amount of water was used for cleaning materials and cooling products. Compositions of discharged wastewater depend on the type of materials. Recently, water used for the manufacturing process has been remarkably reduced. Some households do not even use water for production. A number of households have changed to mainly aluminium melting, a process that does not require cleansing of scraps. Instead, scraps are placed directly into melting pots; molten aluminium is then poured into moulds and left to cool naturally without using water.

According to Dang (2005), materials used for aluminium recycling mainly comprised cans and old/broken pots. Thus, key compositions of wastewater discharged during this stage contained a mixture of dirt from cans. Discharged wastewater from cleaning aluminium scraps contains toxins and oil. Further, based

LIVELIHOOD AND ENVIRONMENT TRADE-OFFS 177

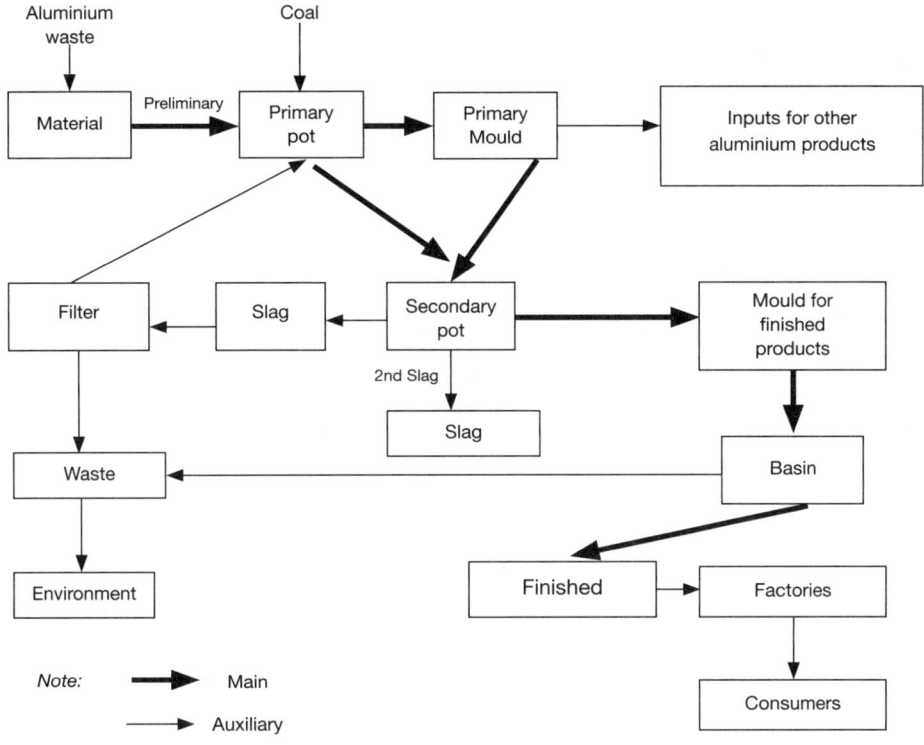

Figure 8.2 *Aluminium production chain*

on survey results, the average amount of water used for production is $1.2m^3$/ton of aluminium products.

A number of products that need surface processing cause wastewater containing some acid, base or chrome compounds. In addition, plating also discharges a great amount of wastewater that causes pollution (wastewater normally has a low pH, containing many metal ions). Cinder filtering also causes wastewater with a great deal of heavy metals. Concentration of manganese, nickel and zinc in wastewater after cleaning finished products is much higher in concentration compared with those in wastewater discharged in the process of cinder filtering. On the other hand, aluminium concentration in wastewater discharged from cinder filtering is higher compared to that from the cleaning.

At present, recycling of other types of scraps, such as lead and zinc, has also started in the village. A number of households are not only engaged in trading scraps for aluminium manufacturing but also other waste materials from broken machines/equipment, electric cables, used electronic materials and batteries, among others. After being sorted out they are then sold to manufacturing households or elsewhere outside the commune. Water used for the cleaning of waste materials/scraps would afterwards contain oil, chemicals and dust, therefore accumulating into high concentrations of oil and toxic chemical pollution. At the same time, waste from aluminium melting has been inappropriately discharged and stored together with other types of wastes/scraps all over the village. Since the waste is not properly sealed, it is swept away by rainwater, thus seriously polluting the village's water resources.

The waste produced from manufacturing activities mainly includes cinder, broken metal scraps and gases from kilns (Dang, 2005). Gases have abundant toxic components such as aluminium gas, COx, SOx and NOx. Such substances as SOx and NOx emitted into the atmosphere would turn into nitric acid (HNO_3) and then permeate the ground and seep into lakes, ponds, rivers and streams. This is the acid deposition process. Consequently, in addition to direct impacts on human health, these gases are main causes of acid deposition phenomenon, adversely affecting water resources.[3]

LIVELIHOOD VERSUS ENVIRONMENT AND HEALTH

Metal craft occupations have brought livelihood security and a certain level of prosperity to Van Mon villagers. However, along with the economic benefits have emerged negative environmental and health impacts that threaten the livelihood benefits of the people and their community.

Livelihood and income stability

Results of the field survey, conducted by the authors, on the valuable possessions of households in the village suggests a certain level of prosperity by rural and semi-rural standards of living in Vietnam. Two out of 10 wealthy households and 3 out of 16 upper middle households sampled have trucks. One out of 10 wealthy households, 2 out of 16 upper middle households and all households among the poor have cars.[4] Respondents in the middle group are the only households that did not own any truck or car. All of the sampled households had motorbikes, TV sets, videos, refrigerators, radios, furnaces, electric fans and gas stoves.

The indebtedness profile of households in the village similarly suggests relative stability of income sources. Based on our survey in 2006, of households that borrowed cash during that year or had been in debt in the previous year, the majority of households in the rich, upper-middle, middle groups and all households in the poor group were able to pay back their loans and were debt free (see Table 8.2). The table also shows that rich households borrowed the highest amount of money (VND 16.5 million per household), followed by the upper-middle (VND 14.3 million per household) and then by the middle (VND 2.1 million per household). In general, however, no household in any of the four groups borrowed more than their annual income. In addition, households did not find it difficult to pay back their debts.

Table 8.2 *Household borrowing in the surveyed year*

Categories	Rich	Upper middle	Middle	Poor
Number of households who borrowed in the surveyed year	8 (80%)	14 (88%)	17 (90%)	0
Loan / household (million VND)	16.5	14.3	2.1	0

Source: field research, 2006

Table 8.3 *Net cash income sources of different social groups of households/ year/capita in 2006 (VND millions)*

	Rich	Upper-middle	Middle	Poor
Metal melting	7.5	6.4	9.3	5.0
Junk trade	6.5	7.8	9.9	8.0
Metal melting & junk trade	0	5.8	0	0
Other	0	0	5.0	0
Total	14.0	20.0	24.2	13.0

Out of a total of 474 households in Man Xa, 235 (almost 50 per cent) are engaged in aluminium and metal smelting. Based on the sample surveyed, the major significance of metal melting and junk trade in the livelihood portfolio of households in all socio-economic groups is undoubted. Table 8.3 reveals some interesting facts.

Unlike other agricultural communes in the Red River Delta the majority of villagers, except for those in the middle households in the craft village of Man Xa, do not earn income from sales of agricultural produce or animal husbandry. Households in the middle group are also those that earned most from high-return occupations – aluminium and metal melting, and junk trading. They are followed by the wealthy households, which earn from metal melting, by the upper-middle and then by the poor respectively in terms of earnings from metal-related occupations. The poor households earn the next most from junk trade, followed by the middle and then by the rich. The upper-middle is the only group that has households engaged in both aluminium and metal melting, and junk trade. No group of households draws income from state wage.

The data shows some further interesting patterns. Unexpectedly, the wealthy households are not the ones who earn the most from all sources.[5] The middle group households are the ones that earn the most from all sources, followed by the upper-middle. By engaging themselves in non-farm sources of income and in both aluminium and metal melting and junk trade the middle and the upper-middle have optimized their earnings compared with the wealthy and the poor households. The data also demonstrates that incomes from non-farming activities are much higher compared with the meagre income earned by a rice farmer (VND 5 million per capita per year). This also explains why despite the majority of people interviewed being quite aware of worsening water pollution in their community, they were still willing to continue and expand aluminium and metal melting.

Water resources degradation

The water resources in Van Mon are common property. This means that previously, and even now, people have common access to water resources. More importantly, villagers do not have to pay for the water they take from wells. Villagers do not care about saving the water at all, since it does not matter how much they use. Prior to *doi moi* reforms, the entire community effectively managed the water resources. Roughly 20 years ago up to 95 per cent of households used water from dug wells of 10–20m deep. However, in the last 10 years, villagers switched to using water from drilled wells with an average depth of 40–45m. This is due to the realization that the quality of the water that they had been using had worsened, and thus they needed to dig deeper in order to obtain better water quality. In the last three years households were found to have drilled wells of 90–95m in depth.

Furthermore, according to elderly people in the village, some 20 years ago there were still seven or eight fish ponds, where water quality was still good enough for raising fish despite the fact that these ponds also served as wastewater receivers. Today, there are only three of these ponds left, one of which is heavily polluted. Some 15 years ago, according to informants, the village had a common well too, which was 7m wide with good water quality. This well was filled in five years ago due to the bad quality of water.

At this point, one might wonder why water quality was still generally good when the villagers were already engaged in aluminium melting. This is because while local households around two decades or so ago had already engaged in aluminium melting their numbers were small. (Indeed, the village's overall population was small.) Now the population has dramatically increased. More importantly, due to an increased market demand, many households today have been intensively engaged in the craft industry. They also carry out not only aluminium, but also zinc and lead melting, which further intensifies water pollution.

In Man Xa, large volumes of wastewater are increasingly discharged into the environment. It is a typical densely populated craft village. Large amounts of water are used for domestic activities and animal husbandry purposes and discharged without any treatment. Likewise, large volumes of industrial wastewater are also discharged directly into local water bodies without treatment.

No drainage canals in the village have met the requirements of hygienic conditions. By the time the research was being carried out most of the sewage canals did not have a cover.

Drinking water

Households in Man Xa village have no access to piped water. Local people rely on water from drilled wells or rainwater for drinking and domestic use. They do not use water filters at all. The demand for clean water in the area is some 60l/person/day and night (DONRE, 2005). The average volume of water exploited is approximately 0.5m^3/day/well. For drilled wells, the figure is higher, which is about 1.5m^3/well/day. Rainwater is also one of the water sources that villagers like to use.[6]

Based on water samples collected from both drilled wells and collected rainwater, the current status of the quality of water used for drinking and daily activities exhibit the following characteristics:

1 The concentration of most heavy metals (except for iron) has not exceeded the standard for drinking water; however, iron concentrations in two drilled well water samples are three to eight times higher than the standard level for drinking water
2 Microorganism parameters (total coliform), on the other hand, are lower than the standard levels
3 Only one out of the three water samples was found with H_2S in a drilled well (however, it is still within the permissible levels)
4 Two samples of drilled well water were polluted by oil and the concentration in one of the two samples exceeded standard levels
5 The concentration of NH_4^+ is high not only in underground water but also in rainwater, exceeding the standard level for drinking water[7]
6 The concentration of BOD_5 in all the water samples is higher, exceeding the standard levels by 3–5 times.

During the rainy season, the concentration of most parameters is lower compared with the dry season. Yet, some parameters such as oil, total nitrogen and H_2S are a little higher than those in dry season. It is noteworthy that oil has been found in the rainwater samples. Nevertheless, the concentration of oil is still at a permissible level (0.008mg/l). Oil was found in all collected samples (drilled well water and rainwater samples). In addition, a number of parameters of BOD_5 and NH_4^+ in the rainwater samples are lower in the rainy season when compared with the dry season, but still exceed the permissible limits. Meanwhile, there is no difference in these parameters in the drilled-well water samples between those taken during rainy season and during dry season.

During the rainy season, the aluminium concentration in the rainwater of households that are not engaged in aluminium melting is quite high (0.9mg/l), exceeding the permissible limits for potable water (0.5mg/l). This can be explained by the fact that aluminium dust created during the manufacturing process scatters and falls on to house roofs and ultimately drains into rainwater tanks. Therefore, these households receive secondary effects even though they are not involved in the melting of aluminium. Meanwhile, those that are engaged in aluminium melting have lower aluminium concentration. It is likely that these households are more careful in the reserving of the rainwater and protecting it from being polluted by aluminium dust.

For almost all of the other remaining parameters, no distinct difference in rainwater samples between aluminium manufacturing households and non-manufacturing ones has been observed. There is also no significant difference in these parameters in the drilled-well water samples and tap water samples. It should be noted that three out of four potable water samples (including treated piped water sample) is polluted by lead with a concentration, that is exceeding the permissible level (0.012–0.016mg/l compared with the standard level of 0.01mg/l). Attention should be paid to piped water management, since some parameters including lead in the treated tap water sample are slightly higher than the permissible level. In summary, Man Xa villagers' potable water of rain and well water show strong signs of being polluted as shown in a number of parameters of BOD_5, NH_4^+ and lead. As a result, it is of great urgency to treat potable water resources to ensure that villagers have access to clean water.

Irrigation canal water and river water

Water samples from rice paddy and irrigation canals collected during the rainy season contain some heavy metal concentrations (like zinc, lead, mercury and chromium concentration) (see Table 8.4 below). But these are much lower than the permissible level for irrigation water (TCVN, 2000).

The Ngu Huyen Khue River provides an important water source for many socio-economic activities in the region, including for irrigation purposes. The Ngu Huyen Khe River flows through Chau Khe Commune (Tu Son District, Bac Ninh province), where a well-known iron manufacturing village named Da Hoi and other artisanal villages like Dong Ky, Phu Lam paper manufacturing village (Tu Son District) and Phong Khe paper manufacturing village (Yen Phong District) are located. Most toxic substances emanating from these villages are discharged directly into the river, which then pours into the Van An water gate in Van An Commune, Yen Phong District, before finally flowing into the Cau river. Untreated sewage is discharged into the environment, canals, ditches, fields and the Ngu Huyen Khue River.

Results of a survey of the quality of water at the Van Mon bridge in 2004 (DONRE, 2005) shows that during the dry season, the quality of water was

Table 8.4 *Selected heavy metal concentration in rice paddy and irrigation canals*

Parameter	Unit of measure	Rice paddy	Irrigation canal
Zinc	mg/l	<0.01	<0.01
Lead	mg/l	0.066	0.003
Mercury	mg/l	0.0001	0.0002
Chromium	mg/l	<0.001	<0.001

not good and the concentration of polluted substances was higher. Results of water sample analysis from the researchers own fieldwork in 2006 confirmed a worse situation. Concentrations of BOD_5, NH_4^+, lead and copper exceeded the permissible level. During the dry season, the BOD_5 concentration exceeded the permissible level though not by much (16.5mg/l compared with the standard level of 10mg/l). However, during the rainy season, the BOD_5 concentration increased dramatically in both river water samples – four times higher than the standard for natural water bodies, despite the water current in the river being stronger during this time. It is possible that during this time the river water becomes not only polluted largely by local waste sources, but also by sewage discharged from upstream of the river.

Further, there are signs of heavy metal pollution like lead in the water samples taken in the rainy season (0.013–0.075mg/l), which are higher than the standard for natural water bodies, but lower than the standard for irrigation water. The copper concentration in most of the water samples is also higher than, or as high as, the permissible level for natural water bodies.

In summary, the water quality of irrigation canals, rice paddies and the Ngu Huyen Khe River is still good enough for agricultural and irrigation purposes. Nevertheless, in terms of natural water, the concentration of some substances exceeds the permissible level. Effective measures to control and treat the quality of the river water are needed to conserve the quality of river water for a wider range of purposes.

Drainage canals and sewage ponds

The concentration of certain substances in the water samples taken from sewage canals is quite high.[8] The concentration of COD varies from 238mg/l to 291mg/l, which is two or three times higher than sewage standard levels of Type B. The BOD_5 concentration is also very high – higher than the standard level of Type B and two out of which exceeded the standard level of Type B. The concentration of ammonia is also very high, which is 51–90ml (50–90 times higher than the standard level of Type B (sewage used for aquaculture and irrigation purposes only) and 5–9 times higher than the standard level of Type C, which must be licensed by the authorities). Phosphorus total concentration is also higher than the permissible level though not by very much. Generally, the metal concentration is the highest. Most are lower than the standard level of Type A (except for iron, which exceeds the standard level of Type B). The oil concentration in drainage canals is higher than the standard level of Type B and C used for industrial sewage. One sample is six times higher than the standard level of Type C.

The village sewage ponds used to be ponds for agricultural purposes and daily activities. However, they are now seriously polluted. The metal concentration such as mercury, lead and copper exceeds the permissible level stipulated for natural water bodies. The concentration of mercury and lead also exceeds the standard level of Type A. The parameters of organic pollution is very high, even higher than the standard level of Type A (BOD_5) and Type B (COD) stipulated for industrial sewage.

The result of the water samples collected from drainage canals and sewage ponds during the rainy season shows that the concentration of substances in water samples taken in the rainy season is very high, even higher if compared with that in the dry season. The concentration of pollutant substances in water samples taken from the sewerage canal of non-manufacturing households is also lower than that of manufacturing ones. However, the concentration of almost all pollutant substances is very high, exceeding the permissible level for household sewage and industrial sewage of Type B.

It can thus be said that sewage canals of manufacturing households are seriously polluted even in the rainy season. The COD concentration approximately exceeds the permissible level regulated for industrial sewage of Type C; the BOD_5 concentration is 2.5 to 2.7 times higher than the permissible level stipulated for industrial sewage of Type C. The NH_4^+ content is also very high (eight times higher than the permissible level regulated for industrial sewage of Type C). These sewage samples contain some metals whose concentration exceeds the permissible level of Type A.

The pollution level in the sewage ponds in the rainy season is lower than that in the dry season. However, the organic pollution level is still high, exceeding the standard level of Type B stipulated for industrial sewage and much higher than the standard level for natural water bodies (i.e. washing, aquaculture and irrigation purposes). Water in those sewage ponds is not safe for any other purpose. Moreover, the accumulation of dirt and toxics must be taken into consideration in the region's sewage management. Similarly, special attention should be paid to the management and the control of the pollution level in sewage canals. The situation calls for necessary measures to treat household sewage as well as production sewage before it is discharged into the village' drainage systems.

Impact on villagers' health, especially on women and girls

Due to lack of environmental awareness among many of the villagers, rubbish is thrown into the drainage canals, thus reducing their draining capacity. Furthermore, organic wastes when degraded pollute the water. Out of 474 households, only 280 households have hygienic toilets. The remaining households did not meet the requirements of sanitation. This is also a major source of water pollution. Stagnant wastewater in the drainage system has caused unpleasant odours. During the rainy season, because the drainage systems do not function well, wastewater spills over from the canals into the village lanes, polluting the environment and posing risks to human health.

Man Xa's rapid development of craft industry, including associated demographic growth, and the absence thus far of solutions to effectively abate water pollution, have combined to make the level of environmental pollution in the village a health hazard. All wastewater receiving water bodies are polluted at an alarming rate, thus posing risks to the environment and to local villagers' health.

To date, no technical research has been carried out to directly establish and explain the causal relationship between degraded water quality in the village and the health status of the local population. However, indicators and signals strongly suggest the negative impact of wastewater on humans in Van Mon. This has caused significant concern among the villagers.

In assessing possible impacts of wastewater on villagers' health, 80 household heads were asked which water resources they used for what purposes. They were also asked to describe the quality of water resources. Further, they were asked whether any members of their household had contracted a water-related disease, and if their answer is positive, whether they had sought medical treatment.

The majority of those interviewed said that Man Xa's water resources are polluted. However, the impact of water pollution on villagers' health is claimed to be insignificant. Further, heads of 47 households were asked: 'During the past 12 months have members of your household suffered from any diseases or health problems?' Figure 8.3 shows that the majority of the rich, upper-middle and the middle claim not to have any health problems. However, a minority in these groups as well as the poor group admit to suffering from colds and respiratory problems.

However, an inspection of the Van Mon's Health Clinic's records shows a different picture. Man Xa villagers' visits to the health clinic have been increasing annually (see Table 8.5). In addition, because local villagers have gone elsewhere to

Table 8.5 *Man Xa villagers' visits to Van Mon Health Clinic*

Year	2001	2002	2003	2004	2005
Visits	547	598	625	670	725

Source: Van Mon People's Committee, 2006

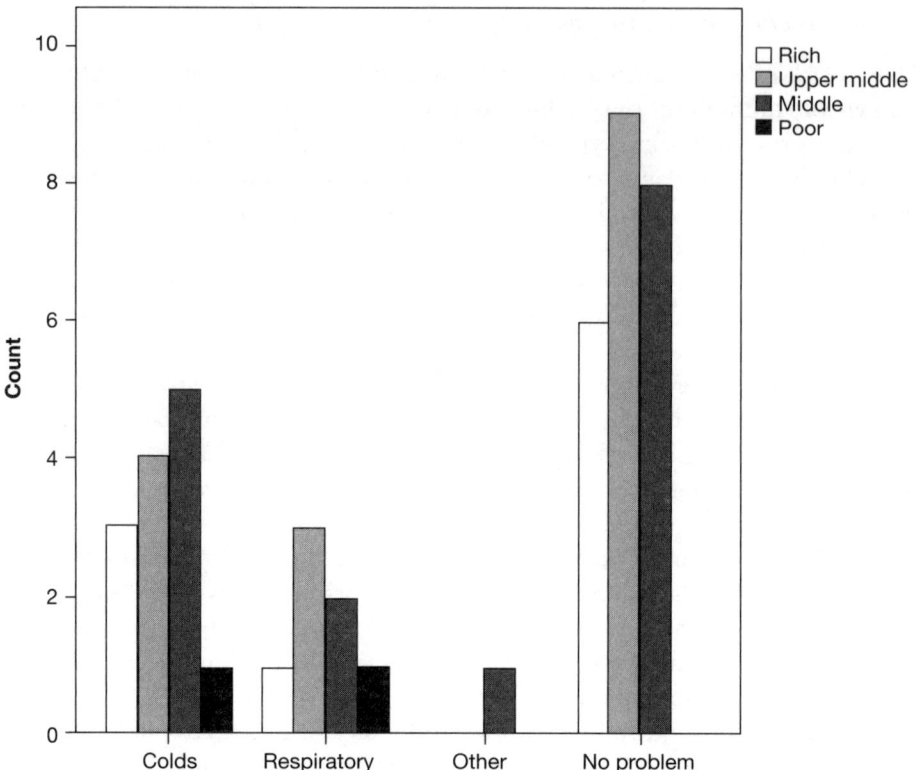

Figure 8.3 *Diseases or health problems suffered by sampled households in 2006*

seek medical treatment, the real number of households who have sought medical treatment is believed to be higher than the figures shown.

The number of deaths in the village increased significantly between 2004 and 2005 (see Table 8.6). The cases of death while giving birth and of infants in 2005 were two to three times higher than the preceding year. The number of people contracting diarrhoea also increased by 200, the majority being children. While the data below presents only two consecutive annual periods and is not sufficient to present a time-series trend, still the increases are significant and require urgent further investigation.

Table 8.7 shows the number of deaths due to cancer for the last four years. It was reported that the majority of those who died of cancer were between 50 and 60 years of age and the majority of them were men (22) and the rest were women (15). The causes of cancer have not yet been specified. However, the number of those who had cancer had increased year by year (based on interviews with the head of the Van Mon Health Clinic). One likely causal explanation posited by local health specialists is the worsening environmental pollution in Van Mon. What lends plausibility to this is the fact that the number of those who died of

Table 8.6 *Diseases and health problems in Man Xa 2004–2005*

STT	Type of disease	2004	2005
1	Number of deaths while giving birth	2	8
2	Number of common deaths	43	76
	At home	36	40
	Elsewhere	7	36
3	Death of infants	5	10
4	Miscarriage	3	5
5	Diarrhoea	426	601
6	TB	713	748
7	Food poisoning	5	15

Source: Van Mon People's Committee, 2006

Table 8.7 *Number of cancer-related deaths*

Type of cancer	2003	2004	2005	2006*
Stomach/intestinal	2	3	4	2
Lung	4	3	6	1
Other	3	4	3	2
Total	9	10	13	5

Note: *2006 data is up to June.
Source: Van Mon People's Committee, 2006

cancer and of those who contracted other diseases in Van Mon was two to three times higher compared to the number of other communes in the district that are not engaged in artisanal activities (Yen Phong District's Health Center records). According to public health specialists, polluted water is the main reason for stomach and intestinal cancer.

Further, interviews with the head of the Van Mon Health Clinic reveal alarming health information. The number of villagers who had visited the health clinic for medical treatment was 1.3–1.5 times higher than those of other communes. The majority of those who had pollution-related respiratory diseases were children (75 per cent). Deaths of expectant mothers and infants were mainly caused by premature birth. In 2004 there were two cases of embryonic death and four cases in 2005.

When the head of the Van Mon Health Clinic was asked about common wastewater-related diseases, she said that these were skin diseases such as rashes and allergy. According to her, three times more people in Van Mon suffered from these diseases compared to other communes. However, the commune's heath clinic did not have the complete supporting data, since villagers did not always come to the clinic for medical treatment for this disease. Instead, they bought medicine and then applied it themselves.

The head of the Van Mon's clinic also believes that stomach cancer is being caused by local water pollution. She recalls that in 2004 eight people, three of whom were children under 10 years, contracted chemical poisoning. Their households were engaged in the trading of junk that had chemicals, which were not carefully checked before being sorted. It was reported that eight people were sent to a hospital in Hanoi where they spent two weeks before making a full recovery. In 2005, there were two more cases of chemical poisoning. This demonstrates that scraps contained many toxic chemicals that were not carefully checked by villagers: when they are washed, toxic chemicals, such as acid, mercury, herbicides and other toxins, pollute local water resources. Further, according to the same informant, two families in Man Xa had babies who died when they were more than one year old. It was reported that they were affected by toxics and even dioxin in the junk. However, no scientific study and evidence has verified this incident.

There is no evidence so far of any significant gender differentiated health impact of local water pollution or bad water quality. However, because of their distinct role and exposure to certain craft activities women have become likely victims in Man Xa in craft-occupational accidents. According to key informants, during the last three years there have been five cases of explosions among workers melting aluminium; the majority were female workers. As a result five women in their late 20s and early 30s lost their sight. Others had their thighs injured or burned. It is important to note that none of these women had health insurance. Further, they had not been provided with protective goggles or work clothing. The burden to support the family is now placed on the husband's shoulders.

In summary, there are strong indications and evidence of water pollution in Man Xa adversely impacting on villagers' health. The number of people who have died of cancer and have pollution-related respiratory and intestinal diseases is increasing. Women and children have been most adversely affected. Although there is no scientific evidence to show that women have been more affected by water pollution or bad quality of living and drinking water they have become victims of the development of artisanal activities in the village. Since villagers' awareness is still low – and perhaps too because of their own material stakes in the craft production – they have not realized nor highlighted the impact of water pollution. But records of the clinic and health experts' opinions strongly point to the contrary.

Discussion: Externalities and the Public–Private Interface

The case of the Man Xa craft village mirrors the major tensions between private sector development and environment and health issues unfolding in the context of Vietnam's current chosen path of rapid industrialization and economic growth. While such tension is also present in the operations of private corporate industries (as well as state-owned industries) in the country, the pervasive, diffused and virtually intractable characteristics of household-level manufacturing in craft villages creates particular and unique difficulties in solving problems through the use of simple state-centric or top-down official planning and regulatory policy instruments.

At the core of the tension between private household craft production and its environmental costs is the right of people to pursue their livelihoods and the notion of natural water bodies as common goods. In the case discussed here, for example, the Ngu Huyen Khue River and all ponds in the village, which provides an important water source for many socio-economic activities in the region (including irrigation purposes, canals and other ponds in the commune), is considered a common property. Open access (e.g. common property) bears no obligation or utility to anyone to manage the water resource, commonly expressed in the popular Vietnamese metaphor – 'No one cries for the common father.' Moreover, no one is fined or punished for polluting the waterways. Villagers therefore discharge waste directly into this water body, which is the cheapest way to dispose of waste. Perhaps they do care that their behaviour would in turn adversely affect their health, but certain pragmatic considerations such as, for example, food security, may blur urgency and importance of their health. The farmers' mentality and the lifestyle of small producers pushes them to look at the short-term profits and ignore or de-prioritize the longer-term and wider benefits and persistent effects of pollution.

Enforcement of environmental law is weak and, more importantly, there are overlaps and no clear mandates between relevant ministries and branches and between ministries and branches and the localities. It is also very doubtful if simple command-and-control legal instruments and enforcement techniques can be effective in tackling pollution of private household-size and micro enterprises. In Vietnam, if enforcement of environmental laws and regulations is already a major problem with regard to private corporate firms, a bigger – and perhaps an insurmountable – problem is millions of household-scale craft production units spread over a wide area and operating with a large dose of informality.

That there are no easy government-imposed solutions to the problem was recently highlighted by a failed plan in 2005 made by the Bac Ninh Provincial People's Committee to designate, relocate and concentrate villagers' workshops in a special 35ha area for craft activity of 300 households. Villagers resisted the plan, and the government failed to force them to move their workshops into a new area whose creation would have required them to give up most of their agricultural lands and retain only a small portion amounting to a seventh of their land. A second alternative plan currently under study by the provincial government is to limit the size of a new manufacturing site of the commune to 5–10ha, in order not to dislocate other agricultural lands in use. According to the second plan, formal registration as enterprises at the district level would be required for every household in order to operate in the new manufacturing site; those failing to do so would be banned altogether from household-level craft manufacturing.

In connection with the second plan, a local office of the environment has been set up recently. This office has obliged all producers in craft villages to have environmental licences, which are issued by the provincial Department of Natural

Resources and the Environment. Those failing to do so will not be able to transport scraps and products into or out of the village. More importantly, their clients cannot buy their products if they don't have a licence. According to the villagers in Van Mon, the province has not issued them licences simply because the provincial government wants to move their workshops out of the village (a residential area) to a concentrated production area. This concentrated area does not as yet exist however. While this new hardline policy might force households to comply with the environmental law, on the other hand, the same might force small manufacturing households out of business, leaving the field only for large producers.

While state-centric and command-and-control regulatory measures alone provide little hope to effectively manage the problem, allowing households to devise their own solutions is a non-starter too. Aside from their own resource limitations, as the Man Xa case shows, households as micro-economic agents tend to be trapped in immediate livelihood and economic gain consideration that have thus far blurred or ignored longer-term environment and health trade-offs. Further, the case also shows that economic reforms have opened up opportunities for many, but have not benefited the entire community. Although the gaps in income may not seem excessive, new access to productive resources has laid seeds for continued rapid social differentiation in the future. The village is stratified and response to market demands has been different for individuals, while recent experiences of transition economies confirm that socio-economic inequality is quite unavoidable in market reforms, excessive inequality and exclusionary development, of course, is undesirable and can be a major block, among others, to achieving environmental sustainability goals. Thus, growth-driven social stratification in the context of liberalization of the economy and market reforms may also encourage atomized, individualistic orientations, and raise the hurdle to realize collective inter-household and community cooperative consensus and actions for managing the public good, a necessary factor to effectively address the pollution in craft villages.

In Vietnam reforms have undoubtedly set in motion an unstoppable momentum for local communities and households to enhance their productivity through private household-level production. To a large extent, these market-based forces have pulled many households out of poverty, especially in the rural and peri-urban areas, and started them on a track of upward social mobility. But, at the same time, privatized household craft production has also created new and complex problems of environment and health impacts. These challenges call for new governance approaches and mechanisms, which are not yet in place. In a new approach to governance, strong state regulation and effective enforcement is indispensable, which Vietnam is still lacking in the field of environmental management. But more importantly, inter-household and community cooperation is important for achieving minimum compliance to statutory regulations and standards. Moreover, household and community cooperation and voluntarism are necessary to draw up responsive and enforceable environmental plans and standards at the local level. Increased social capital at this level could also facilitate locally appropriate and innovative institutional arrangements and technological developments. For example, decentralized wastewater management proven to be effective in many peri-urban areas in developing countries involves decentralized decision-making and participatory planning and water segregation at the source (that is, at the household level). The same holds true for effective community-based health programmes that emphasize prevention, protection and early monitoring of pollution-related diseases.

Thus far, however, household and community roles in environmental governance in Vietnam have been largely insufficient. Yet, an overly pessimistic outlook in this regard may also be unfounded. Certain scholars have noted in recent studies (Dang, 2009; Nguyen, 2009) that not only in Van Mon but also elsewhere in Vietnam, health impacts are often the catalyst for local people to consider water as a common-pool resource that needs to be managed collectively and effectively for the sake of improving local health conditions.

Conclusion

This case has demonstrated how weak governance has produced a combination of environmental and health impacts on the same communities that exploit local water resources. Future governance innovations in this handcraft production context would require effective state actions and instruments at both the central and local levels. Perhaps more importantly, there is a need for mechanisms that enable community and households to engage with the public authorities in collaborative and participatory planning and negotiated decision-making to handle the tension between household livelihood interests, community welfare and ecological sustainability and macro-economic goals pursued by the state. The *doi moi* policy changes allow individual households freedom to enhance their well-being through individualized production, which has lifted many out of poverty. This is a major achievement in and of itself. However, the economic benefits have been accompanied by adverse environmental and social changes, and the current challenge is to energize and mobilize a cooperative and community ethos in genuine partnership with the state, to take care of the environmental and health costs of the *doi moi policy* in peri-urban and rural localities.

Acknowledgements

The authors wish to thank the CIDA-UEMA Project, which supported the recently concluded action research and policy advocacy initiative on craft villages in Vietnam.

Notes

1 LEP, among others, also requires environmental impact assessments (EIAs) of new industrial plants and land-use plans, authorizes the government to levy environmental taxes and charges, and the official issuance of ambient and source standards for various types of pollution and the necessary inspections to pursue compliance to these standards (DiGregorio et al, 2003, p195).
2 Man Xa started its silk weaving as far as Y Lan's time during the Ly Thanh Tong dynasty (1023–1072). At the time, villagers made long dresses and brassieres. In addition, they were engaged in alcohol and agricultural production and making tofu. But from 1958 fabric weaving fell into oblivion.
3 Chemical abbreviations:
  NH_4^+ ammonium
  BOD_5 biochemical oxygen demand
  COD chemical oxygen demand
  CO_x carbon oxides
  H_2S hydrogen sulphide
  HNO_3 nitric acid
  NO_x nitrogen oxides
  SO_x sulphur oxides
4 The stratification of households into four groups – rich, upper-middle, middle and poor – was done using primary survey information that included housing quality indicators, household's material consumption goods and income. This is based on a random sampling of 47 households out of 447 of Van Mon's total household population.
5 Although a caveat to this is that heads of the rich households might not have correctly reported their incomes in our survey.
6 A few households, which are located near Van Mon's People's Committee (including the commune's health clinic), have been supplied with tap water, a provision of the programme Clean Water for Rural Areas.
7 Contrary to common practice in the area, rainwater thus needs treatment before use.
8 Due to the absence of specific standards for artisanal villages' sewage, we have used instead standards for industrial sewage – the Vietnamese standard 5945/1995 that is higher than the standards for household sewage – as the benchmark.

References

Ba, N.X. (1997) 'Handicrafts could pave the way to industrialization', *Vietnam News*, p4
Cuong, D.M. and Nguyen, Q.H. (1998) 'Rural-Rural Spontaneous migration: Problems and Solutions: Institute for Labor and Social Issues', in proceedings of the International Seminar on Internal Migration: Implications for Migration Policy in Vietnam, UNDP, Population Council & Ministry of Agriculture and Rural Development, Hanoi
Dang, K.C. (2005) 'Scientific base and practice research for construction of policies and methods for settling environmental issues in Vietnam's craft villages', Summary Report on Science and Technique, Ministry of Natural Resources and Environment, Hanoi

Dang, K.C. (2009) 'Vietnam's craft villages and environment', policy paper of the project 'Policy Advocacy Campaign in Vietnam: Stakeholders, Wastewater Management and Air Pollution Treatment in Craft Villages in the Red River Delta of Vietnam', Center for Natural Resources and Environmental Studies (CRES), Vietnam National University, Hanoi

DONRE (Department of Natural Resource and Environment)/Bac Ninh province (2005) 'Report on current status of the environmental of Bac Ninh', Bac Ninh province

DiGregorio, M., Tien, T., Lan, H., Ha, N. and Nguyen, T.L.A. (1999) 'The environment of development in industrializing craft villages', working paper, Center for Natural Resources and Environmental Studies, Vietnam National University, Hanoi

DiGregorio, M., Rambo, A.T. and Yanagisawa, M. (2003) 'Clean, green and beautiful: Environment and development under the renovation economy', in H.V. Luong (ed.) *Postwar Vietnam. Dynamics of a Transforming Society*, ISEAS and Rowman and Littlefield Publishers, Inc., UK and Singapore, pp171–199

Gainsborough, M. (2004) 'Key issues in the political economy of post-*doi moi* Vietnam', in E. MacCargo (ed.) *Rethinking Vietnam*, Routlege, London and New York, pp40–52

Giao, H. and Cuong, H. (1995) 'Vietnam's private economy in the process of renovation', in I. Norlund, C. Gates and V. Dam (eds) *Vietnam in a Changing World*, Curzon Press, UK, pp151–158

Ha, N., Kant, S. and Maclaren, V. (2008) 'Shadow prices of environmental outputs and production efficiency of household-level paper recycling units in Vietnam', *Ecological Economics*, vol 65, pp98–110

Jamieson, Neil, Le Trong, Cu and Rambo, A. Terry (1998) 'The development crisis in Vietnam's mountains', East West Center Special Reports, No 6, Hawaii

Kelly, P., Lien, T., Hien, H., Ninh, N. and Adger, W. (2001) 'Managing environmental change in Vietnam', in W. Adger, P. Kelly and N. Ninh (eds) *Living with Environmental Change*, Routledge, London and New York, pp35–58

Kerkvliet, B. (2005) *The Power of Everyday Politics. How Vietnamese Peasants Transformed National Policy*, ISEAS, Singapore

Luong, H. (2003) 'Introduction', in H. Luong (ed.) *Post War Vietnam – Dynamics of a Transforming Society*, ISEAS, Singapore

Minh, N. (2002) 'Foreign direct investment-led development for better urban environmental management: The case of Hanoi, Vietnam', unpublished PhD dissertation, Urban Environmental Management Field of Study, Asian Institute of Technology

MONRE (Ministry of Natural Resources and Environment) (2008) 'National report on the environment in Vietnam's craft villages', Hanoi

Muller, H. (2005) 'Private-sector development in a transition economy: The case of Vietnam', *Development in Practice*, vol 15, nos 3&4, pp349–361

NEA/WB/DANIDA (National Environment Agency/World Bank/Danish Agency for International Development) (2002) Vietnam Environment Monitor

Nguyen, T.L.A. (2009) 'Environmental Pollution and Villagers' Health', paper presented at the workshop on Minimizing Environmental Pollution in Craft Villages in Vietnam in Van Mon Commune, Yen Phong District, Bac Ninh province on 29 May 2009

O'Rourke, D. (2002) 'Community-driven regulation: Toward an improved model of environmental regulation in Vietnam', in P. Evans (ed.) *Livable Cities? Urban Struggles for Livelihood and Sustainability*, University of California Press, Berkeley and Los Angeles

O'Rourke, D. (2004) *Community-Driven Regulation: Balancing Development and the Environment in Vietnam*, The MIT Press, Cambridge, Massachusetts

People's Committee of Van Mon Commune (2006) 'Tong ket viec thuc hien nhiem vu phat trien kinh te-xa hoi, quoc phong an ninh nam 2006' (Report on the Implementation of the 2006 Plans on Socio-economic Development and Social Security in Van Mon)

Sajor, E. and Thu, M. (2009) 'Institutional and development issues in integrated water resource management of Saigon River', *The Journal of Environment and Development*, vol 18, no 3, pp268–290

Sinh, B. (2004) 'Institutional challenges for sustainable development in Vietnam: The case of the coal mining sector', in M. Beresford and A. Tran Ngoc (eds) *Reaching for the Dream: Challenges of Sustainable Development in Vietnam*, ISEAS, Singapore

TCVN (2000) 'Water Quality', General Department of Quality Measurement, Ministry of Science, Technology and Environment, TCVN/TC, p147

Van Mon People's Committee (2006) 'Bao cao benh an cua Tram Y te xa Van Mon' (Report on Villagers' Health Situation of Van Mon's Health Clinic), Van Mon Health Clinic

VCG (Vietnam Consultative Group) (2005) 'Governance, Vietnam development report 2005', Joint Donor Report to the Vietnam Consultative Meeting, 1–3 December 2004, Hanoi

VDR (2008) 'Vietnam development report 2008: Social protection', Joint Donor Report to the Consultative Group Meeting

VietNamNet (2007) Hộ kinh doanh cá thể (KDCT) có vai trò quan trọng trong việc tham gia vào phát triển kinh tế, song chưa được chú ý hỗ trợ (Registered Households Play Important Roles in Economic Development, but not yet Paid Due Attention), Hanoi

World Bank (1999) 'Attacking poverty', Vietnam Development Report 2000, Poverty Reduction and Economic Management Unit, East Asia and Pacific Region, Washington

Part IV

Climate Change and the Rights of the Vulnerable

9

Climate Change in the Asian Highlands: Socio-economic Implications for the Mekong Region

Jianchu Xu

INTRODUCTION

The 'Asian highlands' refers to the mountainous area that includes southwest China (Yunnan province, part of Sichuan and eastern Tibetan Plateau) together with northern mainland Southeast Asia, lying 300–3000m above sea level (masl) and occupying about one-quarter of Asia's land surface. The Asian highlands are the 'Water Towers' of Asia[1] being the source of 11 major rivers: Syr Darya, Amu Darya, Indus, Ganges, Brahmaputra, Irrawaddy, Salween, Mekong, Yangtze, Yellow and the Tarim (Xu et al, 2007; 2009) (see Figure 9.1).

The Asian highlands like other mountain areas in the world are being affected by global warming especially in the areas above 3000masl[2] that comprise a series of parallel and converging ranges forming the highest mountain regions in the world, where over 140 million people live (ICIMOD, WWF & Ramsar, 2004). These high-altitude ecosystems contain the most extensive areas of glaciers and permafrost (ICIMOD, WWF & Ramsar, 2003) outside high latitudes, in other words the Earth's surface nearest either pole, especially the part within either the Arctic or the Antarctic Circle.

Climate change is an issue of global importance and poses a number of known and predicted impacts in the Asian water towers (Immerzeel et al, 2010). The direct and indirect impacts of alterations in the rainfall regime, vulnerability to disasters such as landslides and floods and geographic shifts in infectious diseases are just a few of the most pressing concerns. In parallel, societal pressures on water resources are increasingly caused by the demands of rapidly growing economies in the Mekong region.

The Mekong region looks set for major changes due to climate change, making it one of the most vulnerable places on Earth (WWF, 2009). Biodiversity is important for ensuring food security and local livelihoods of the 300 million people living in the Mekong region. For example, if there is a one metre rise in sea level in Vietnam, 70 per cent of the Delta will be inundated with salt water resulting in a loss of over 2 million ha of rice, and nine key biodiversity areas in the Mekong delta will be negatively affected (Carew-Reid, 2008; WWF, 2009). The implications for Vietnam as a major rice-producing country are clear.

However, the wider effects on the whole of the Mekong region while being felt are still not completely known. There are various lines of thinking at the national level, with regard to how countries will adopt either a prevention or adaptation approach in dealing with the impending reality of climate change impacts. There has also been an increasingly compelling line of analysis rooted in issues of local well-being as well as rights of affected people (see Chapter 10, this volume), considering options for how communities can mobilize indigenous social and ecological knowledge and the resources to deal with the uncertainty of climate change, while at the same time creating ways to link into sources of support and innovation at the national and regional as well as international levels.

This chapter looks at the implications of climate change in the Asian highlands and growing socio-economic pressures on resources that, in combination, are set to pose far-reaching impacts for the diverse ecosystems and societies in the Mekong region. This chapter argues for more inclusive governance processes to address these complex and multi-layered impacts, both climate- and development-related, within the various socio-political and ecological landscapes in the Mekong region.

Climate change in the Asian Highlands

The Asian highlands contribute a run-off of up to 60 per cent of the total volume of water to the river valleys (Bandyopadhyay et al, 1997). According to Viviroli and Weingartner (2002) mountains in general provide a) disproportionately large amounts of discharge; b) seasonal retention of discharge through the accumulation of snow and ice; and c) reliable amounts of run-off due to the regularity of the melting process and the storage capacity of glaciers.

The Asian highlands consisting of the Tibetan Plateau and inner mountain ranges cover a total area of $3,846,131 km^2$ at elevations above 3000masl (Li et al, 2009). Water is often stored as glaciers, ice and snow both temporally and permanently. The area provides approximately 9 million cubic metres of freshwater annually. It has been estimated that about 30 per cent of the water resources of

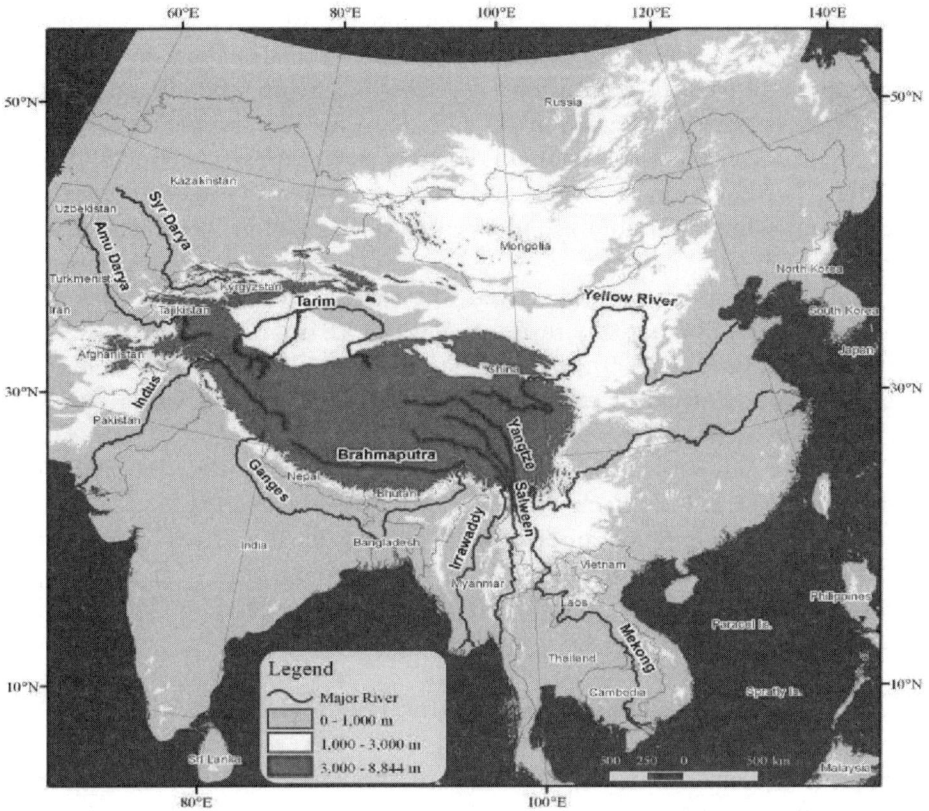

Figure 9.1 *The Asian highlands comprising the Tibetan Plateau, mountain ranges, the main rivers and river basins*

Source: Xu et al, 2009

the eastern Himalayas are derived from the melting glaciers, snow and ice: this proportion increases to nearly 50 per cent in the Indus in the western Himalayas and becomes as high as 80 per cent in the upper reaches of the Tarim Basin. The complexity and intensity of high altitude wetlands, lakes and river network systems enable them to hold water for fairly long periods of time. Melting glaciers, ice and snow replenish freshwater significantly in early spring and summer. Highlands are often a combination of hilly and flat plateau areas covered with alpine vegetation, bush and grassland, which soak up storm flows and capture nutrients and sediments, protecting water quality in downstream rivers and lakes. The highlands serve as a water tower by compensating for the deficiency in rainfall and snowmelt during dry and drought years and alleviating the variability in supply of water to stabilize river flows.

Climate change is a key concern in the high elevation areas since the temperature increase in these areas is higher than the global average (IPCC, 2007). A major concern is the impact on glacial water storage and downstream water discharge and availability (IPCC, 2007). Key ecological effects of climate change include shifts in precipitation patterns, glacial retreat, permafrost melting, floods and drought, soil erosion, ecosystem degradation and desertification. Key social effects of climate change include changes to livelihood practices and risks associated with natural disasters. Thereby, climate change adds to environmental and socio-economic stresses in causing uncertain use and access to water resources (Ives and Messerli, 1989).

Studies suggest that glacier-fed rivers will swell for a few decades as glaciers and snow melt, but will eventually turn into a trickle (Barnett et al, 2005; Rees and Collins 2006; Xu et al, 2009). In the long term, glaciers are predicted to disappear with major consequences for the seasonal water availability in the Asian river basins including the Mekong River.

Land-use and land cover changes can have additional effects on hydrology compounding and interacting with climate-related changes (Ma et al, 2009). Hence, future precipitation and discharge patterns in Asian river basins remain uncertain, but the social and political responses are even greater, with prospects for water potentially being a trigger for both conflict and cooperation (Miller, 2008).

OBSERVED AND PROJECTED CHANGES

Changes have already begun to be felt in the Asian highlands and further downstream in the Mekong region. This section provides critical information detailing the observed and projected changes on temperature and participation and the complex responses and impacts on water resources.

Temperature

Average temperature increases in the Asian highlands ranged from less than 1° to 3°C in the last century (the global average is 0.74°C) (IPCC, 2007). Over the last century, warming in Nepal and the Tibetan Plateau has increased progressively with elevation (Liu and Hou, 1998; Shrestha et al, 1999). Warming at high altitudes (>3500masl) was 0.25°C per decade between 1961 and 1990 (Liu and Hou, 1998), approximately three times the global rate. The increase in winter temperatures is greatest on the Tibetan Plateau (0.32–0.33°C per decade), mainly because average minimum temperatures have increased; this warming accounts for the biggest contribution to increases in average annual temperatures in the highlands (Niu et al, 2004).

Due to its high altitude and topographic diversity, the Tibetan Plateau is very sensitive to global climate change (Liu and Chen, 2000) In the Karakoram and Hindu Kush mountains, winter mean and maximum temperatures have increased significantly, while summer mean and minimum temperatures have declined

(Archer and Fowler, 2004); the same applies to the eastern Himalayas in southwest China. Ma et al (2009) analysed annual temperatures between 1965 and 2005 and found that in 1986, mountain ranges in the upper Salween basin experienced a temperature regime shift. This coincides with analysis from the Tarim River Basin (Chen et al, 2003). The length of the growing season (daily temperature >10°C) on the Tibetan Plateau has increased by 15 days during the last three decades. The IPCC's Fourth Assessment Report (AR4) predicted that the average annual mean warming would be about 3°C by the 2050s and about 5°C in the 2080s over the Asian land mass (IPCC, 2007), with temperatures on the Tibetan Plateau rising even more (Rupa Kumar et al, 2006). In the Mekong region, temperatures will increase at a rate of 0.02 to 0.03°C per year to 2050, with higher rates of warming in the upper Mekong including Yunnan province and the Tibetan Plateau (Johnston et al, 2010).

Precipitation

Research on precipitation trends in recent decades in the Asian highland points to a general increase – on average 3.4mm/decade – mostly due to an increasing trend in winter precipitation, while summer rainfall has seen a slight decrease (Wilkes, 2008). Although some researchers refer to an overall trend of warming and drying on the Tibetan Plateau, in fact compared to the relatively uniform trends in temperature across the Plateau, there has been much more geographical variation in precipitation trends (Xu et al, 2009). During the last few decades, inter-seasonal, inter-annual and spatial variability in rainfall has been observed across the Asian highlands. The highlands have experienced both increasing and decreasing precipitation trends. Increases have occurred in the northeastern Tibetan Plateau, eastern and central areas and in northern Pakistan (Xu et al, 2007; Ma et al, 2009). Nepal has shown no long-term precipitation trend between 1948 and 1994 (Shrestha et al, 2000). The western Tibetan Plateau has exhibited a decreasing precipitation trend. Ma et al (2009) noted a trend towards early arrival of the monsoon, decreased rainfall during mid-monsoon and increased rainfall in the late monsoon in the eastern Himalayas from 1965–2006. Monsoon patterns have shifted, but the picture is ambiguous because most studies exclude Himalayan precipitation processes due to topographic complexity (Shrestha et al, 2000). Another feature of climate change is the increasing frequency and magnitude of extreme weather events such as intense rainfall, typhoons and droughts, which have substantial impacts on local economies and human lives in Asian highlands in general and the Mekong region in particular (Ma et al, 2009; WWF, 2009). There is also increasing unevenness of rainfall distribution in space; in most cases wet areas have become wetter and dry zones have become drier (Xu et al, 2009, Johnston et al, 2010).

Complex responses and effects

The changes in atmosphere, which are largely driven by human-industrial activities in the bio-geophysical environment, have direct impacts on the cryosphere at high altitudes and both directly and indirectly impact the biosphere and human society through affecting the hydrological processes in mountain ecosystems. The broad predictions of global climate change, especially the emphasis on shifts in mean temperature, do not take into account important regional complexities in the highlands, which are related to the influence of topography and elevation. If climate change mainly involves vertical shifts in precipitation and thermal conditions, ruggedness, elevation and orientation will also modify the significance of regional and local climate changes. The highest mountains, or those facing or funneling the prevailing winds, may retain substantial, if diminished, glacial cover, whereas lower or less favourably oriented watersheds may lose theirs. Furthermore, most climate models predict intensification of the Asian monsoon. On a regional scale this could result in an increase in precipitation, although local effects are poorly understood. Moreover, climate change means not only warmer temperatures, but also changes in precipitation, evapotranspiration, soil and air moisture, run-off and river flow as well as groundwater through hydrological cycles.

Climate change is expected to accelerate water cycles and thereby increase the available, renewable freshwater resources (Oki and Kanae, 2006). Temperature changes have a predominantly regional character, whereas precipitation changes are more locally determined and very difficult to analyse and to predict, especially in mountain areas and river basins (Jian et al, 2006). Equally critical are issues related to the structure, processes and resilience of ecosystems or biosphere and human adaptations to them; bearing in mind those ecosystems and humans are possibly already stressed by adaptation to topoclimatic diversity. In general, local impacts of the climate do not follow single or simple paths, whether in terms of plant ecology, stream hydrology, erosion and sedimentation, extreme events or human activities.

Although studies show marked variations in the local impacts of climate change, such as orographic precipitation in different valleys and at different elevations within the same mountain range, most of the region remains unstudied in terms of a baseline for assessment or prediction of these complexities.

The consequence, rate and magnitude of effects vary in different zones, which include vegetation shift, frequent wildfires, changes in freshwater supplies and problems to human health (Xu et al, 2009). Table 9.1 depicts the potential effects of climate changes on water resources, agriculture, biodiversity and human health in different critical zones from the highlands to the uplands, from the lowland to coastal and urban areas for example in the Mekong region. The general circulation model (GCM) shows that the highlands have experienced a decrease in freshwater run-off in most arid and semi-arid drylands and an increase in the eastern Himalayas; an increase in forest cover and wildfire frequency have been experienced in most highlands of Asia (Scholze et al, 2006). Furthermore, increasing human activities in the highlands through land-use and land-cover change, infrastructure development and tourism have exacerbated the vicious cycle of climate change.

Impacts on water resources

Climate change presents very serious risks to freshwater resources. Climate change may result in increasing temperatures, decreasing snow packs and earlier snow melting, and that will certainly reduce the flow of water in the rivers originating from the highlands, particularly in the dry season.

The glaciers in Asian highlands are retreating faster than the world average (Dyurgerov and Meier, 2005) and thinning at the rate of 0.3–1m/year. Besides glacial retreat, impacts of climate change include disappearance of small wetlands including lakes. Glacial melt provides freshwater particularly in the arid areas of western China and during the critical dry season period until the monsoons begin. The supply of freshwater, or meltwater from snow and ice, in large river basins is projected to increase over the following decades as perennial snow and ice decrease (Xu et al, 2009). Later, however, most scenarios suggest a continual decrease by the 2030s and leading to catastrophic proportions by the 2050s.

Shi (2001) predicts that small glaciers (less than 2 km^2) will be more sensitive to global warming; meltwater will reach its peak value at present, and will decrease or even disappear by 2050. Medium-sized glaciers of 5–30km^2 will reach their meltwater peak value by 2050. The larger glaciers (areas exceeding 100km^2) will retreat slowly. In the short run, animal husbandry and agriculture could benefit from temperature rise and increases in meltwater discharge as long as good water management practices and proper irrigation facilities are introduced, particularly in dryland and arid areas. However, effective and efficient water management technology is still lacking in many areas especially to cope with long-term decreases in water supplies including the severe drought in the Mekong region in 2010 (Qiu, 2010).

Natural water-induced risks

Due to environmental fragility and sensitivity to climate change of the Asian highlands, shifts in rainfall patterns and temperatures can induce extreme weather events such as flash floods and droughts, river floods, landslides and debris flows, snow avalanches and even wildfires when there is insufficient rainfall (Chapter 10, this volume). In the Upper Mekong or the southeastern Tibetan Plateau where there is heavy precipitation and the temperature is higher than in other areas, the rising temperatures could accelerate the ablation and retreat of glaciers, potentially resulting in more frequent flash floods and landslides in upland watersheds.

Complexities arise, especially from interactions among different cold climate elements: freeze-thaw and peri-glacial processes, snowfall, valley wind systems, avalanches, glacial processes and seasonal or spatial balance between frozen and liquid precipitation, albedo and evaporation. Not only are they likely to change with general climate shifts, but also interactions among them can buffer, exaggerate or redirect the impacts of change in any one element. The most rapid and varied interactions occur through the 'vertical cascade' between different topoclimates – zones stacked vertically and on slopes of differing orientation – notably transport of moisture, run-off, sediment and dissolved solids downslope. The occurrence and impacts of major hazards, such as avalanches, debris flows, landslides and flash floods, also have a bearing on downslope, down-glacier and downstream cascades. Whereas snow avalanches and glacial lake outburst floods (GLOFs) predominate at very high elevations (>3000m), landslides, debris flows and landslide dam outburst floods (LDOFs) are more common in the middle mountains (1000–3000m). Riverine floods are the principal hazards in the lower valleys and plains; the causes of these floods are related to climatic conditions (Chalise and Khanal, 2001; Dixit, 2003; Xu and Rana, 2005).

Societal Use of Resources

Climate change is posing a number of risks and impacts as outlined above. This section looks at how the Asian highlands are facing additional pressures that are being compounded by climate change. For example, pressures due to securing access to viable land, water and other resources due to the rapidly growing economies in the Mekong region resulting especially in the expansion of water-diversion infrastructure and cash crop and tree plantation areas in the lowland. The construction of dams for storing water and demarcation for uplands for watershed conservation in turn redistributes water resources, denies access to some and raises political tensions among different stakeholders in the uplands and lowlands. Highland populations are also looking to intensify land use and cash cropping, which will need more water and fertilizer inputs (Kahrl et al, 2009). It also results in less water flows to the lowlands as well as decreased water quality

Land-use change with implications for water use

Over 80 per cent of the population in the Asian highlands depend either on full- or part-time farming for their livelihoods (Thulachan, 2001). The pace, magnitude and spatial reach of land-cover and land-use changes in the Asian region have increased over the last half century as a result of land reclamation, for example, for rubber plantations in Yunnan and northern Laos (Ziegler et al, 2009). Degradation of grasslands on the Tibetan Plateau is attributed to overgrazing and climate change (Wilkes, 2008). Land-use changes affect fauna and flora, which contribute to local, regional and global climate changes; and are the primary source of soil, water and land degradation (Pielke, 2005; Sthiannopkao et al, 2007; Ma et al, 2009).

Land-use decisions are also water-related decisions (Falkenmark, 1999). For example, Ma et al (2009) found that forestation reduces stream flow due to increased evapotranspiration, but that tree planting could mitigate impacts of climate change on water resources through regulating extreme rainfall events. The rapid emergence of water-dependent rubber is the hallmark of a larger land-cover transition that has been sweeping through montane mainland Southeast Asia in recent decades (Ziegler et al, 2009). The conversion of secondary forests to rubber threatens biodiversity, results in reduced total carbon biomass and has negative hydrological consequences. Altering ecosystem services from the Asian highlands affects the ability of biological systems to support human needs in the Asian region. The provisions people obtain from ecosystems (e.g. food and water), cultural services (e.g. spiritual and recreational benefits) and support services (e.g. pollination, nutrient cycling, and productivity) are likely to be altered. The regulating services (e.g. predator-prey relationships and flood and disease control) are expected to change significantly. For example the increasing frequency in extreme weather events (e.g. droughts, floods and typhoons) are affecting the production of grains, fish supply and forests (WWF, 2009). The quantity and quality of water resources is shifting. There is a change in conditions resulting in increases in the spread of invasive species and diseases such as malaria and dengue fever, which are becoming more prevalent. Endemic morbidity and mortality due to diarrhoea primarily associated with floods and droughts can be expected due to disruption of the hydrological cycle (WWF, 2009). If the Asian monsoon substantially changes, as mentioned above, the people in the Mekong region are likely to face situations of food insecurity.

While climate warming reduces the storage of freshwater in glaciers, land-use changes have contributed greatly to wetland losses in western China. China has lost $127km^3$ of water storage capacity as a result of wetland losses in the western plateau over the past 50 years (Wang et al, 2006).

Besides climate factors, land management determines the quality and flow of water resources. Land-use practices are inextricably linked with water resources. Land-use and land cover changes are intrinsically linked with the hydrological cycle and their impacts have been studied in depth for decades (Ma et al, 2009; 2010). There are two types of land use activities that have a fundamental impact on livelihoods:

1 land use for irrigation drainage and flood protection, or, for example, limitations imposed by water availability on biomass production. This type of land use is called 'water-dependent' land use;
2 land use that has an impact on rainwater partitioning through soil and vegetation or impacts related to the function of water as a carrier of solutes and silt in the landscape. This type of land use is called 'water-impacting' land use.

Land-use impacts on hydrological parameters and sediment transport are inversely related to the spatial scale on which the impacts can be observed (FAO, 2002). In contrast, impacts of land-use changes on water quality parameters may be relevant on the meso- and macro-scales. It is important to note that the impact of these land-use changes is variable in terms of time-scale. While the quality of water in rivers and lakes can be restored in quite short time, the biodiversity destroyed will take several thousands of years to recover to its original condition.

Agricultural intensification and water pollution

Asian society is agriculture based and water-dependent. Moreover, Asian per capita water use is far above the world's average. The Asian and Pacific region is home to more than 60 per cent of the world's population and agricultural land. The region is also the largest consumer of water by far, with a withdrawal rate of 2384 billion m^3 per year, which is more than the consumption of rest of the world. The Asian, and more specifically the Mekong region, uses the largest proportion of its water – nearly 79 per cent of its total withdrawal – for agriculture (www.asiawater.org/water.html, last accessed 4 March 2010). Water use is expected to increase rapidly together with population growth and agricultural production, particularly in South Asia, in comparison with Southeast Asia including the Mekong region.

Extraction of water for irrigation has, unavoidably, a big impact on river flows. Agricultural intensification has also resulted in dramatic increase in fertilizer use. Nitrogen-based fertilizer is one of major sources of greenhouse gas (GHG) emissions in China (Kahrl et al, 2009). High use of N-fertilizers also leads to N$_2$O emissions through the nitrification-denitrification process (Kahrl et al, forthcoming). Levels of N-fertilizer use in Asian countries are high both on a per yield and a per area basis relative to countries with similar agricultural profiles (Novotny et al, 2009).

IMPLICATIONS FOR THE MEKONG REGION

Given the known and predicted climate change impacts in a region already facing multiple pressures and demand on resources, a range of potential effects of climate change are expected on ecosystems and society in the Mekong region (see Table 9.1).

The highland Mekong

The highlands of the Mekong region are often termed Montane Mainland Southeast Asia (MMSEA) and include the uplands of Yunnan of Southwest China, Myanmar, Thailand, Lao PDR, Vietnam and part of Cambodia. The countries in the MMSEA share a number of ethnic, historical, cultural and biophysical features

Table 9.1 *Possible effects of climate change on ecosystems and society from mountain top to delta in the Mekong region*

Critical zones	Water	Agriculture	Biodiversity	Livelihoods and health
High plateaus	Warmer winters and more rainfall, snow avalanches, melting glaciers and potential glacial lake outburst floods; increase in water levels in highland lakes in the short run and decreases in the long run	Early spring might cause more overgrazing, degradation of rangeland, desertification, snow storms and invasive species.	Woody vegetation moves upward, increase in the number of endangered alpine species, weedy species may spread; ecosystem will deteriorate due to increasing rodent population	Cholera and diarrhoea increase. Increase in new diseases such as Avian flu due to interaction of wildlife, livestock and human beings.
Up and watersheds	Flash floods, landslides, debris flows, and landslide dam outburst floods occurring more frequently; increase in run-off during monsoon, low flow and droughts will decrease during the dry season, increase in soil erosion and sediment transport downstream; silt in the run-off will contaminate water supplies and clog hydroelectricity plants	Higher carbon dioxide levels and water temperatures may increase grain yields, irregular monsoon patterns will delay rice planting and harvests although rice yields may increase during good years	Wildfires and pests increase, woody vegetation increases, wildlife moves upwards to high altitudes, changes in composition of biodiversity due to different performances of species to climate change	Fire haze increases air pollution and emerging disease risk, cholera and diarrhoea increase; schistosomiasis will move into higher wetlands and lakes. Malaria and dengue fever become more widespread in the uplands.
Lowland plains	Changing rainfall patterns, decrease in freshwater supplies (run-off) (in river basins), severe droughts, decrease in ground water levels, damage to wetland from reduction in water availability	Rice yields will decline as temperatures increase changes to cropping patterns and productivity, Farms will be vulnerable to increasing pests and natural disasters (e.g. floods and droughts)	Biome change from forest to non-forest, increase in loss of agricultural diversity and invasive species both in water bodies and the ecosystem generally; acceleration of forest degradation; loss of wet and dry forest ecosystems	Heat waves will kill more people in the lowland plains, poor sanitation together with environmental poisoning will cause an increase in health issues; higher prevalence of infectious diseases
Coastal areas (e.g. delta)	Stronger cyclones, intrusion of saltwater into water supplies, decrease in groundwater, urban floods; sea levels could rise by up to one metre	Salination of farmlands, warmer water will threaten fish farms	Aquatic biodiversity decreases due to salination, warmer water and water pollution	Heat-related illnesses as well as dengue fever, cholera and water- and sewage-related diseases will increase along with respiratory problems related to urban pollution and increasing migration; inundations of coastal zones affecting local communities; displacement and migration of lowland and coastal peoples

such as settlement patterns, land use, ecological landscapes, livelihood activities and associated ideological and cosmological elements. The highland Mekong is well-known for its high cultural and ecological diversity. Although there is a high occurrence of natural hazards, the highlands have been productive and prosperous historically because of their microclimates, which made a diversity of non-timber forest products possible.

Many ethnic communities in the Mekong highlands have historically suffered political marginalization (although some have a degree of political autonomy in some parts of the region), first during the nation-building processes and then later economic marginalization in the transition of countries to a market-based economic development system. Many of the highland populations are among the region's poorest and most politically disadvantaged. Living far from the centres of commerce and power, they have little influence over the policies and decisions that influence their lives and contribute to the deterioration of mountain environments.

The impacts of climate change are hardly new but the challenges arise from the speed at which they are occurring, in combination with continuing pressures on land, water and other resources. The highland people have, however, been affected severely by drought in the region due to destruction of forest cover and a history of poor water management (Qiu, 2010). These challenges are pressing highland populations to respond and adapt to changes both at a rapid and in innovative ways. Highland people are adapting to water-related stress for example by making adjustments in their livelihood practices, institutional arrangements and social relations. Reducing water-induced risks and coping with water stresses are fundamental parts of the local knowledge, innovations and practices of the highland populations and in particular ethnic communities (Xu and Rana, 2005).

The water Mekong

The water Mekong evolved originally from the then-Mekong Committee, established in 1957, later becoming the Mekong River Commission (MRC), which was established under the Mekong water sharing agreement signed in 1995 among the governments of the lower Mekong basin. The agreement between Cambodia, Lao PDR, Thailand and Vietnam is for the sustainable use and development of the water resources in the lower Mekong basin. Key provisions of the Mekong Agreement apply to the main stem of the river, not to tributaries, and only to the four downstream riparian countries. Under the current agreement, there is no provision or agreement for upper riparian countries such as China to share water and discuss their plans for building more dams on the main stem of the Mekong River. However, the agreement could be relevant for the entire basin and accommodate the interests of China, should they wish to join in the future (Dore et al, 2010).

In 2010, in response to lower riparian states' concerns on the drought in the Mekong basin and the likely impact of the dams in the Upper Mekong River,

China invited representatives of the MRC governments to visit the Xiaowan and Jinghong dams on the Lancang Jiang (Upper Mekong River).

Climate change has impacts on agriculture both directly at the local level due to changes in temperature and rainfall, and at the regional level, through changes in water regimes (Johnston et al, 2010). Climate change projections for the Mekong River Basin to 2030 suggest that overall annual precipitation, run-off and flooding could increase because of higher wet season rainfall (Eastham et al, 2008). Shifting interactions among different water uses – hydropower, irrigation and fisheries uses – will be a major challenge for water resources management and interacts strongly with climate induced changes (Costa-Cabral et al, 2007). Both floods and droughts are becoming more frequent and of greater magnitude. There are many ongoing initiatives in the Mekong region to address climate change. For example, the MRC launched a Climate Change and Adaptation Initiative in 2009 to establish a Mekong Panel on Climate Change (MRC, 2009), which would meet every three years to analyse and report on the state of climate change and adaptation efforts. Also, an Adaptation Forum has been organized by the Regional Climate Change Adaptation Knowledge Platform for Asia and the Asia Pacific Adaptation Network in response to the demand for effective mechanisms for sharing information on climate change adaptation and developing adaptive capacities in Asian countries. The Asian Institute of Technology/United Nations Environment Programme Regional Resource Centre for Asia and the Pacific, the Stockholm Environment Institute, the Swedish Environmental Secretariat for Asia and the United Nations Environment Programme (UNEP) have jointly established the Forum.

Climate change in the Mekong, as it poses increased uncertainty and insecurities, has become a key issue in intergovernmental negotiations over water allocation agreements and disaster management protocols (Xu et al, 2009).[3]

The economic Mekong

The economic Mekong or the Greater Mekong Sub-region (GMS) as it is known under economic development programs supported by the Asian Development Bank (ADB) is the grouping of Cambodia, Laos, Thailand, Vietnam and Myanmar together with Yunnan province in China. The GMS does not include the entire area of the Mekong River Basin, and yet it does include extensive areas located in other river basins along the coastal areas. GMS member states provide the political context for decisions about resource and economic policies, and their perspectives on highland regions are reflected in the outcome (Thomas et al, 2008). China's emergence in the last decade as a major global economic power has had a profound effect on the landscape and waterscape in the Mekong region, particularly in increasing demand for natural resources including agricultural products. The volume of agricultural products that China imports from its neighbours, and the values of its investment in plantations, will directly shape agricultural production and water use in the Mekong region (Johnston et al, 2010).

This type of rapid economic growth has only been possible as the Mekong region countries restructured their economies away from a primary focus on agriculture into greater emphasis on industrial and service sectors, along with increased global economic integration. Economic growth plays a key role in drivers of change in water availability. Urbanization, cash crop and timber plantations and expansion of industries all contribute to rising demand for water. As economic changes have also penetrated the previously remote highlands, it has brought increasing commercialization of agriculture and multiple cropping seasons where previously there were only one or two crops per year. This has been causing higher water demands especially for cropping in the dry season, as well as greater pollution from fertilizers of rivers and waterways.

In order to facilitate greater commercial and market-based production, there has been a rapid expansion and upgrading of transportation and water storage infrastructure. Large dams and other infrastructure interventions in the Mekong and its tributaries rivers are being driven by the growing demand for electricity to fuel the region's economic and industrial development. Rapid economic growth is accompanied by increasing energy demand, especially in China, Thailand and Vietnam. The storage capacity of planned dams in Mekong River Basin is in the order of 50,000mm^3 (World Bank, 2004).

There are growing concerns, however, about the possible negative impacts of the large dams in terms of their economic viability as well as impacts on ecology, ethnocultural heritage and geophysical processes in the region. The mainstream and tributary dams in the upper Mekong pose aggregate implications for the natural flood regimes in the lower Mekong basin, especially for the Tonle Sap Lake in Cambodia and the Mekong Delta in Vietnam, crucial for maintaining rice and fish productivity and the proper functioning of wetlands (Molle et al, 2009). Climate change together with land-use changes are probably further aggravating sediment loads in the rivers (Lu et al, 2010), therefore posing reductions to the lifespan of dams. The most recent feasibility designs of the Lower Mekong mainstream dams prepared by private sector developers have not taken climate change explicitly into account. This is in part due to the lack of project-level methodologies and variations in modelling results and unreliable baseline information, thus increasing the uncertainties of climate change and their effects (ICEM, 2010).

Several projects are in the pipeline at different stages in different countries including along untapped reaches of many rivers. For example, the Salween River has several projects in the pipeline in both China and Burma/Myanmar; Myanmar has signed power sale agreements with Thailand to sell 1500MW of energy produced in the Salween river basin. In the Yangtze basin the technically feasible hydropower potential is about 197,000MW or 52 per cent of China's total potential (Kajander, 2001). In addition a further 23 per cent of China's hydropower potential is in the southwest rivers. Both climate change and economic development has geopolitical implications for water resource use.

Conclusion

The ecological health of the Asian highlands has implications directly for about 1.3 billion people living within 11 river basins and 3 million indirectly, particularly in terms of availability of freshwater supplies. While uncertainties about the rate and magnitude of climate change and its potential impacts in the Asian highlands prevail, there is no question that climate change, combined with continuing societal demands on natural resources for economic development, is gradually and powerfully changing the ecological and socio-economic landscape in the Mekong region, particularly in relation to water resources. It is imperative to revisit and redesign development policies, management and conservation practices, and appropriate technologies as well as research agendas related to water resource management. Adaptation to changing water availability (maintaining water-related services and reducing water-induced risks) intended to cope with climate change in the Mekong region can create opportunities as well as offset the dangers of a warming planet; but they must be identified and adopted ahead of, rather than in reaction to, dangerous trends. Improving access to information on climate change and its ecological effects is essential.

Local communities, in particular in the highlands, have continued to live with and survive natural hazards such as flash floods, avalanches and droughts for millennia. Building local adaptive capacity in the socio-ecological systems in the face of climate change is doubly important and is an important step towards achieving sustainable livelihoods. Supporting local resilience and encouraging strategies to cope with surprises and long-term changes are necessary, unlike earlier notions of improving people's adaptations to relatively stable and known habitats. Climate change, as a public and global issue, has evolved from a narrow interest in the hydro-meteorological sciences to a broad recognition that both the social consequences and policies in response have implications for all aspects of human development. Adaptive policies and major efforts to reverse the human drivers of climate change have to be incorporated into all sectors: balanced investment in water infrastructure and institutions, integrated water and land management, conservation of natural ecosystems including wetland, rangeland and forest, alternative rural energy and a better healthcare system.

There is a need to further explore the building of resilient institutions and adaptation strategies in diverse Mekong landscapes informed by local knowledge, innovations, practices and concerns.

Acknowledgements

This work was funded by: Asia-Pacific Network for Global Environmental Change Research (Grant ARCP2008-15NMY-Nikitina) and the M-POWER governance network project PN 50 of the Challenge Program on Water and Food (CPWF), which was financially supported by the International Fund for Agricultural Development and Echel Eau.

NOTES

1 Demonstrating the importance of mountains for providing freshwater for the downstream areas, the symbolic term 'water towers' for mountains is widely adopted today (Bandyopadhyay et al, 1997; Liniger et al, 1998; Messerli et al, 2004; Viviroli et al, 2007).
2 The highlands can be subdivided into an alpine zone (above 3000masl) and a montane zone (300–3000masl). The term 'Montane Mainland Southeast Asia' (MMSEA) is used to describe the areas in the montane and alpine zones (Thomas et al, 2008).
3 For example, within the Association of Southeast Asian Nations (ASEAN), the 'Singapore Declaration on Climate Change, Energy, and the Environment; Singapore Resolution on Environmental Sustainability and Climate Change' which emphasizes cooperation by forming an ASEAN Working Group on Climate Change, as well as plans to develop an ASEAN Climate Change Initiative and studies to assess the impacts of climate change on the Southeast Asia Region. Within the Greater Mekong Sub-region (GMS), at the Environment Ministers' Meeting held in Vientiane 28–30 January 2008, delegates agreed that sub-regional programmes need to be in place to help GMS countries to better prepare for needed actions against climate change, as well as equip them with the capacities to adhere to international conventions to promote sound environmental management. At the ADB-GMS 15th Ministerial meeting in Cha-Am, Thailand on 19 June 2009, the Ministers from Cambodia, China, Lao PDR, Myanmar, Thailand and Vietnam released a statement prioritizing the reduction of environmental risks to local livelihoods and GMS development plans, including those posed by climate change for the next three years.

REFERENCES

Archer, D.R. and Fowler, H.J. (2004) 'Spatial and temporal variations in precipitation in the Upper Indus Basin, global teleconnections and hydrological implications', *Hydrology and Earth System Science*, vol 8, no 1, pp47–61

Bandyopadhyay, J., Rodda, J.C., Kattelmann, R., Kundzewicz, Z.W. and Kraemer, D. (1997) 'Highland waters – a resource of global significance', in B. Messerli and J.D. Ives (eds) *Mountains of the World, A Global Priority*, Parthenon Publishing Group, New York

Barnett, T.P., Adam, J.C. and Lettenmaier, D.P. (2005) 'Potential impacts of a warming climate on water availability in a snow-dominated region', *Nature*, vol 438, no 17, pp303–309

Carew-Reid, J. (2008) 'Rapid Assessment of the Extent and Impact of Sea Level Rise in Vietnam', Climate Change Discussion Paper 1, International Centre for Environmental Management (ICEM), Brisbane, Australia

Chalise, S.R. and Khanal, N.R. (2001) 'An introduction to climate, hydrology and landslide hazards in the Hindu Kush-Himalayan region', in L. Tianchi, S.R. Chalise and B.N. Upreti (eds) *Landslide Hazard Mitigation in the Hindu Kush-Himalayas*, ICIMOD, Kathmandu, pp51–62

Chen, Y.N., Cui, W.C., Li, W.H. and Zhang, Y.M. (2003) 'Utilization of water resources and ecological protection in the Tarim River', *Acta Geographica Sinica*, vol 58, no 2, pp215–222

Costa-Cabral, M., Richey, J., Goteti, G., Lettenmaier, D., Feldkotter, C. and Snidvongs, A. (2007) 'Landscape structure and use, climate, and water movement in the Mekong River basin', *Hydrological Processes*, vol 22, pp1731–1746

Dixit, A. (2003) 'Flood and vulnerability: Need to rethink flood management', in M.M.Q. Mirza, A. Dixit and A. Nishat (eds) *Flood Problem and Management in South Asia*, Kluwer Academic Publishers, Dordrecht, Boston and London, vol 28, no 1, pp155–179

Dore, J., Robinson, J. and Smith, M. (eds) (2010) *Negotiate – Reaching Agreements Over Water*, IUCN, Gland, Switzerland

Dyurgerov, M.D. and Meier, M.F. (2005) *Glaciers and Changing Earth System: A 2004 Snapshot*, Institute of Arctic and Alpine Research, University of Colorado, Boulder, Colorado, p117

Eastham, J., Mpelaskoka, F., Mainuddin, M., Ticehurst, C., Dyce, P., Hodgson, G., Ali, R. and Kirby, M. (2008) 'Mekong River Basin Water Resources Assessment: Impacts of Climate Change', Water for a Healthy Country National Research Flagship, Australia's Commonwealth Scientific and Industrial Research Organisation

Falkenmark, M. (1999) 'A land-use decision is also a water decision', in M. Falkenmark, L. Andersson, R. Castensson and K. Sundblad (eds) *Water – A Reflection of Land Use, Options for Counteracting Land and Water Mismanagement*, Swedish Natural Science Research Council, Stockholm, pp58–78

FAO (2002) 'Land-water linkages in rural catchments', *FAO Land and Water Bulletin 9*, FAO, Rome

ICEM (2010) 'MRC SEA on the Mekong Mainstream: Impacts Assessment (Opportunities and Risks)', discussion draft for stakeholder meeting, 14 May

ICIMOD, WWF & Ramsar (2003) 'Report of the Regional Workshop on Wetland Conservation and Wise Use in the Himalayan High Mountains (Kathmandu, Nepal)', unpublished

ICIMOD, WWF & Ramsar (2004) 'Wetland Conservation and Wise Use in the Himalayan and Central Asian High Mountains (Sanya, China)', Workshop Report, unpublished

Immerzeel, W.W., Ludovicus, P.H., van Beek, Marc and Bierkens, F.P. (2010) 'Climate change will affect the Asian water towers', *Science*, vol 328, June, pp1382–1385

IPCC (Intergovernmental Panel on Climate Change) (2007) 'Climate change 2007: The physical sciences basis', Summary for policy makers, IPCC, vol 21, IPCC, Geneva

Ives, J.D. and Messerli, B. (1989) *The Himalayan Dilemma: Reconciling Development and Conservation*, John Wiley and Sons, London

Jian. M.Q., Qiao, Y.T., Yuan, Z.J. and Luo, H.B. (2006) 'The impact of atmospheric heat sources over the eastern Tibetan Plateau and the tropical western Pacific on the summer rainfall over the Yangtze-River Basin', *Advances in Atmospheric Sciences*, vol 23, no 1, pp149–155

Johnston, R.M., Hoanh, C.T., Lacombe, G., Noble, A.N., Smakkhtin V., Suhardiman, D., Kam, S.P. and Choo, P.S. (2010) *Rethinking Agriculture in the Great Mekong Subregion: How to Sustainably Meet Food Needs, Enhance Ecosystem Services and Cope with Climate Change*, International Water Management Institute, Colombo, Sri Lanka, p26, doi10.3910/2010.207

Kahrl, F., Tennigkeit, T., Wilkes, A., Xu, J.C., Su, Y.F. and Yan, M. (2009) 'A Pro-Growth Pathway for Reducing Net GHG Emissions in China', ICRAF Working Paper No 93, Beijing

Kahrl, F., Li, Y., Su, Y., Tennigkeit, T., Wilkes, A. and Xu, J.C. (forthcoming) 'Greenhouse gas emissions from nitrogen fertilizer use in China', *Environmental Science and Policy*

Kajander T. (2001) 'Water resources, large dams, and hydropower in Asia', Master's thesis, Helsinki University of Technology, Espoo, Finland

Lebel, L., Xu, J.C., Bastakoti, R. and Lamba, A. (2010) 'Pursuits of adaptiveness in the shared rivers of monsoon Asia', *International Environmental Agreement: Politics, Law and Economics*, vol 10, no 4, pp355–375

Li, Z.Q., Shilpakar, R.L. and Xu, J.C. (2009) 'Mapping high altitude wetlands in Himalaya', unpublished project report, ICIMOD, Kathmandu

Liniger, H.P., Weingartner, R. and Grosejean, M. (1998) 'Mountains of the world: Water towers for the 21st century', *A Contribution to Global Freshwater Management, Mt. Agenda*, Department of Geography, University of Berne, Berne

Liu, X.D. and Hou, P. (1998) 'Relationship between climate warming and altitude in Tibetan-Qinghai Plateau and its adjacent areas', *Plateau Meteorology*, vol 17, no 3, pp245–249

Liu, X.D. and Chen, B.D. (2000) 'Climate warming in the Tibetan Plateau during recent decades', *International Journal of Climatology*, vol 20, pp1729–1742

Lu, X.X., Zhang, S.R. and Xu, J.C. (2010) 'Climate change and sediment flux from the roof of the world, Earth surface processes and landforms', *Earth Surface Processes and Land Forms*, vol 35, no 6, pp732–735

Ma, X., Xu, J.C., Luo, Y., Aggarwal, S.P. and Li, J.T. (2009) 'Response of hydrological processes to land cover and climate change in Kejie Watershed, Southwest China', *Hydrological Processes*, doi10.1002/hyp.7233

Ma, X., Xu, J.C. and van Noordwijk, M. (2010) 'Sensitivity of streamflow from a Himalayan catchment to plausible changes in land-cover and climate', *Hydrological Processes*, doi10.1002/hyp.7602

Messerli, B., Viviroli, D. and Weingartner, R. (2004) 'Mountains of the world: Vulnerable water towers for the 21st Century', *Ambio*, vol 13, pp29–34

Miller, K. (2008) 'Climate change and water resources: The challenges ahead', *Journal of International Affairs*, vol 61, pp35–50

Molle, F., Foran, T. and Kakonen, M. (eds) (2009) *Contested Waterscapes in the Mekong Region: Hydropower, Livelihoods and Governance*, Earthscan, London

MRC (2009) 'Adaptation to climate change in the countries of the Lower Mekong Basin', MRC Management Information Booklet Series, no 1, Vientiane, Laos

Niu, T., Chen, L. and Zhou, Z. (2004) 'The characteristics of climate change over the Tibetan Plateau in the last 40 years and the detection of climatic jumps', *Advances in Atmospheric Sciences*, vol 21, pp193–203

Novotny, V., Wang, X., Englande, A., Bedoyam, D., Promakasikorn, L. and Tirado, R. (2009) 'Comparative assessment of pollution by the use of industrial agricultural fertilizers in four rapidly developing Asian countries', *Environment, Development and Sustainability*, doi10.1007/s10668-009-9207-2

Oki, T. and Kanae, S. (2006) 'Global hydrological cycles and world water resources', *Science*, vol 313, pp1068–1072

Pielke, R.A. (2005) 'Land use and climate change', *Science*, vol 310, no 9, pp1625–1626

Qiu, J. (2010) 'China drought highlights future climate threats', *Nature*, vol 465, no 13, pp142–143

Rees, G.H. and Collins, D.N. (2006) 'Regional differences in response of flow in glacier-fed Himalayan rivers to climate warming', *Hydrological Processes*, vol 20, pp2157–2167

Rupa Kumar, K., Sahai, A.K., Krishna Kumar, K., Patwardhan, S.K., Mishra, P.K., Revadkar, J.V., Kamala, K. and Pant, G.B. (2006) 'High resolution climate change scenario for India for the 21st Century', *Current Science*, vol 90, pp334–345

Scholze, M., Knorr, W., Arnell, N.W. and Prentice, I.C. (2006) 'A climate-change risk analysis for world ecosystems', Proceedings of the National Academy of Sciences, vol 103, no 35, pp13116–13120

Shrestha, A.B., Wake, C.P., Mayewski, P.A. and Dibb, J.E. (1999) 'Maximum Temperature Trends in the Himalaya and its Vicinity: An Analysis Based on Temperature Records from Nepal for the Period 1971–94', *Journal of Climate*, vol 12, pp2775–2786

Shrestha, A.B., Wake, C.P. and Dibb, J.E. (2000) 'Precipitation Fluctuations in the Himalaya and its Vicinity: An Analysis Based on Temperature Records from Nepal', *International Journal of Climate*, vol 20, pp317–327

Shi, Y.F. (2001) 'Estimation of the water resources affected by climate warming and glacier shrinkage before 2050 in West China', *Journal of Glaciology and Geocryology*, vol 23, no 4, pp333–341

Sthiannopkao, S., Takizawa, S., Homewong, J. and Wirojanagud, W. (2007) 'Soil erosion and its impacts on water treatment in the northeastern provinces of Thailand', *Environmental International*, vol 33, no 5, pp706–711

Thomas, D.E., Ekasingh, B., Ekasingh, M., Lebel, L., Hoang, M.H., Ediger, L., Thongmanivong, S., Xu, J.C., Sangchyoswat, C. and Nyberg, Y. (2008) 'Comparative assessment of resource and market access of the poor in upland zones of the Greater Mekong Region', ICRAF, Chiang Mai

Thulachan, P.M. (2001) *State of mountain agriculture in the Hindu Kush-Himalayas: A regional comparative analysis*, ICIMOD, Kathmandu

Viviroli, D. and Weingartner, R. (2002) 'The significance of mountains as sources of the world's freshwater', *GAIA*, vol 11, no 3, pp182–186

Viviroli, D., Durr, H.H., Messerli, B., Meybeck, M. and Weingartner, R. (2007) 'Mountains of the world, water towers for humanity: Typology, mapping, and global significance', *Water Resources Research*, vol 43, W07447 doi10.1029/2006WR005653

Wang, Z.S., Zhou, C., Guan, B.H., Deng, Z.F., Zhi, Y.B. and Liu, Y.H. (2006) 'The headwater loss of the Western Plateau exacerbates China's long thirst', *Ambio*, vol 35, no 5, pp271–272

Wilkes, A. (2008) 'Towards mainstreaming climate change in grassland management policies and practices on the Tibetan Plateau', ICRAF Southeast Asia Working Paper No 67, World Agroforestry Centre – ICRAF China, Beijing

World Bank (2004) 'Modelled observations on development scenarios in the Lower Mekong Basin', *Mekong Water Resources Assistance Strategy*, World Bank, Vientiane, p142

WWF (2009) 'The Greater Mekong and climate change: Biodiversity, ecosystem services and development at risk', Greater Mekong Programme, Bangkok, Thailand

Xu, J.C. and Rana, G.M. (2005) 'Living in the mountains', in T. Jeggle (ed.) *Know Risk*, UN Inter-agency secretariat of the International Strategy for Disaster Reduction, Geneva, pp196–199

Xu, J.C., Shrestha, A., Vaidya, R., Eriksson, M. and Hewitt, K. (2007) 'The melting Himalayas: Regional challenges and local impacts of climate change on mountain ecosystems and livelihoods', ICIMOD Technical Paper, International Centre for Integrated Mountain Development, Kathmandu

Xu, J.C., Grumbine, R E., Shrestha, A., Eriksson, M., Yang, X., Wang, Y. and Wilkes, A. (2009) 'The melting Himalayas: Cascading effects of climate change on water, biodiversity and livelihoods', *Conservation Biology*, vol 23, no 3, pp520–530

Ziegler, A.D., Fox, J.M. and Xu, J.C. (2009) 'The rubber juggernaut', *Science*, vol 324, pp1024–1025

10

Linking Climate Change Risks and Rights of Upland Peoples in the Mekong

Jianchu Xu and Rajesh Daniel

INTRODUCTION

A changed climate is already here as the world faces threats of disaster from more frequent and intense cyclones, heavy rainfall events, sea-level rise and warmer temperatures with adverse effects on crops, ecosystems and human health (IPCC 2007a,b). Much of the climate burden is falling on the poor and marginalized peoples in developing countries, highlighting the disasters and impacts they face, the issues of rights and justice with respect to the allocation of resources as well as burdens and risks (Kates, 2000; Thomas and Twyman, 2005). This chapter draws on the context of the uplands of the Mekong region to highlight the interconnected dimensions of climatic risks and the rights of those affected by climate change and development, in particular social and political rights[1] as enshrined in the International Convention on Economic, Social and Cultural Rights.

The uplands of the Mekong region within Montane Mainland Southeast Asia (MMSEA)[2] (see Figure 10.1) comprise steep slopes and tectonic activities where natural hazards and risks are omnipresent. As a result of the seasonal shifts in monsoon weather patterns, a large part of upland Asia is exposed to increased annual floods and droughts (Bates et al, 2008). Climate change multiplies these risks. Evidence shows that extreme fluctuations of climate such as rapid or sudden shifts in rainfall can result in either too much water causing loss of lives and property from flash floods or too little water leading to drought, loss of crops and death of livestock (Xu et al, 2009).

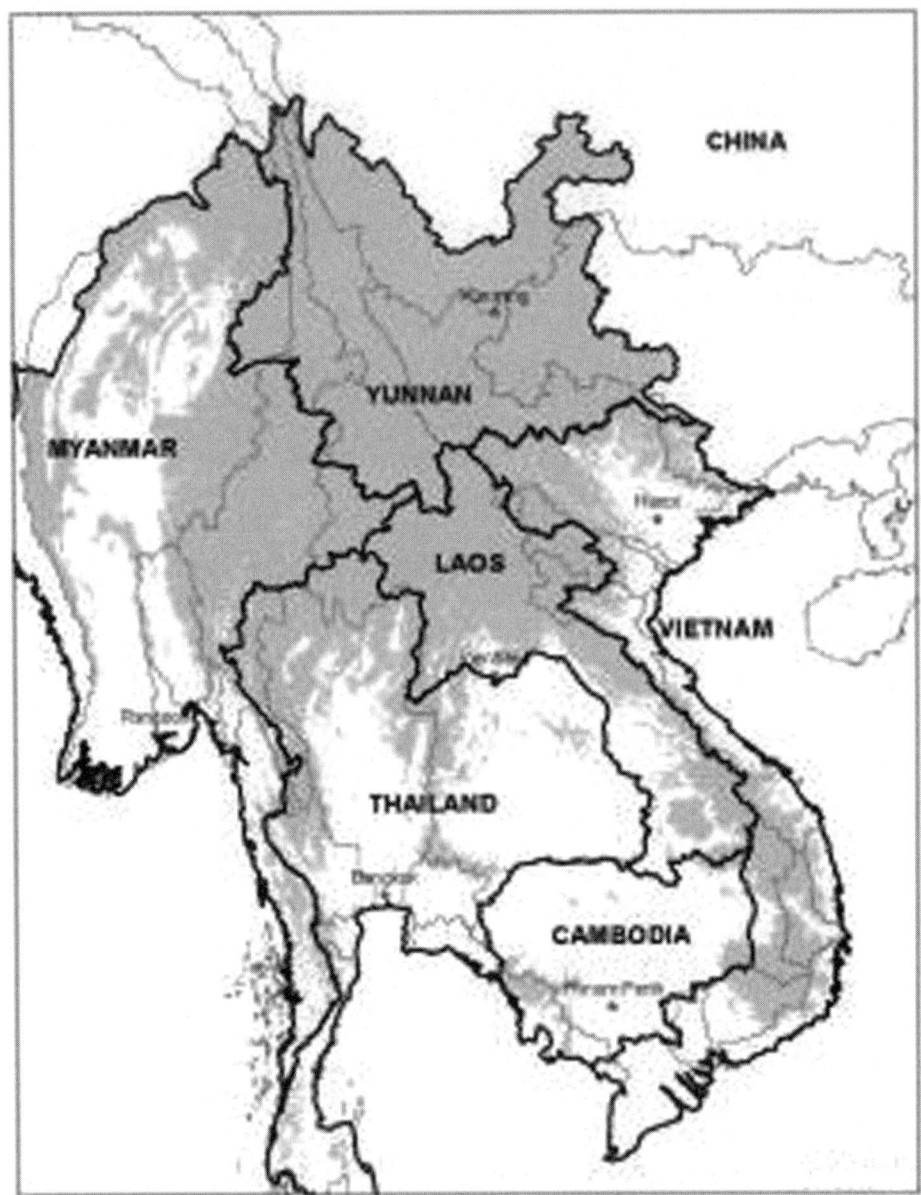

Figure 10.1 *Map of Mekong uplands and Montane Mainland Southeast Asia*

Source: David Thomas (pers. comm., 2009)

Areas shaded in grey are the montane and alpine zones of areas that are 300–3000masl.

Climate impacts are often socially constructed (Ribot, 2009). Apart from climate-induced risks, various government development policies, institutional settings and expansion of regional, national and international markets have also placed upland peoples in a position of greater vulnerability. The rapid pace of regional economic integration has meant that previously subsistence-oriented livelihoods are quickly shifting towards a market-orientation, often with the encouragement of the government. New cash crops are accompanied by new forms of financial management for local people, and debt has become a major concern across the uplands. In this period of transition, the risk of natural disaster has compounded economic implications for local livelihoods.

Response to climate change in terms of mitigation and adaptation can also multiply risks. Structured measures for climate change adaptation such as embankments might redistribute flood risks (Lebel et al, 2007) while hydropower development poses threats to river ecosystems and local livelihoods such as fisheries. Tree plantation projects, for instance, to earn carbon credits are taking away common lands and secondary forests used by upland communities (O'Brien et al, 2007).

Douglas and Wildavsky (1982) define risk as a joint product of knowledge about the future and consent about the most desired prospects. Living with multiple risks, poor and marginalized groups must manage the costs and benefits of overlapping natural, social, political and economic hazards (Xu and Rana, 2005; Ribot, 2009). The rural poor have successfully faced threats linked to climate variability in the past in the form of mobility, storage and communal pooling of water and other natural resources, diversification, architecture and market exchange in rural settings as the basic mechanisms through which households address risks in securing livelihoods (Agrawal, 2009).

The diversity of the Mekong uplands includes multiple livelihoods (from shifting cultivation in the humid tropics to nomadic herding on the high Tibetan plateau, from rice terraces to tea gardens), multiple ethnic cultures (more than 50 officially recognized ethnic nationalities and hundreds of linguistic groups) and numerous vulnerabilities (see Box 10.1).

While the upland peoples in the Mekong region experience both threats and opportunities from climate change or development actions, many people, in particular the economically poorer, face disproportionate vulnerabilities in terms of loss of livelihoods and assets in the face of climate variability and global change (Sen, 1981). This situation is exacerbated by the fact that upland peoples are frequently blamed for environmental risks in downstream and coastal areas, despite the complex and still poorly understood causal linkages between change in the mountains and change in lowland areas. As will be discussed further below, policies to halt perceived environmental degradation in the uplands has often resulted in increased vulnerability for upland people, while the risks to lowland society remain unmitigated.

Climate-related risks will have direct and indirect human rights impacts.[3] Climate change is already undermining the realization of a broad range of internationally protected human rights: right to health and even life; right to food, water, shelter and property including access to natural resources; rights associated with livelihoods and culture, migration and resettlement and with personal security in the event of conflict (UNHCHR, 2009). Multiple risks relating to climate and economic development impact on a range of rights, including participation in decision-making (Molle et al, 2009).

Climate impact analysis that links to human rights can prove useful in formulating detailed policy and research agendas to inform overarching climate change policy options (ICHRP, 2008), including strategies for mitigation and adaptation, and for particular ecological settings such as the uplands in the Mekong region.

Adaptive management to global climate change can often lie beyond the capabilities of upland and indigenous people, even though many communities are dealing with climate risks using their traditional ecological knowledge systems. While recognizing that some groups are more resilient than others, the capabilities of local people and groups can be strengthened when appropriately assisted through partnerships with government and non-government organizations to ensure equitable access to resources and benefits. Yet low 'capacity' for adaptive management is often a product of constraints within the governance system of a country or region. Traditional ecological knowledge and environmental management practices are deeply rooted in local natural and cultural landscapes, but as upland areas are increasingly integrated into lowland social, economic and political systems, the pressures on indigenous and local knowledge increase. However, representation of upland people in national political processes is often constrained to varying degrees across the region. This means that what is perceived as a problem of many may actually be a problem of empowerment: hence the argument for inclusion of rights as a central component in considering risks from climate change.

Our analysis of the climate risks in the Mekong uplands through a rights-based approach tries to address the climate-related vulnerabilities including both natural and human-induced hazards for the upland peoples, in particular for the poorest, in the Mekong region. The chapter's intent is to bring together three threads that in the authors' view are not adequately addressed in the climate change discourse: uplands, poor/marginalized people and their disproportionate risks and the human rights dimension.

The risks-rights framework helps to better understand and address climate-related risks and impacts in the uplands of the Mekong region and improve policies related to upland governance and climate change to benefit the poorer and marginalized peoples of the upland populations. Mapping geophysical hazards and socio-political constructed risks provides an entry point for concerted pro-poor climate change adaptation efforts.

The chapter includes the following sections: the risks analysis of geophysical hazards and climate change in the Mekong uplands; accelerated risks and socially

Box 10.1 The Mekong Uplands, Peoples and Livelihoods

Slope, aspect and altitude determine the fundamental characteristics of mountain habitats. Topographic diversity adds to the small-scale variations in physical environment. Geographically, latitude (distance from the equator), continentality (distance from oceans) and topographic features (direction and altitude) affect climate and local weather patterns, rendering some mountains almost permanently wet, others dry and yet others highly seasonal (Xu and Rana, 2005). Geological conditions add dimensions of diversity and influence soil development, soil type, erosion processes and vegetation cover. As climate varies according to altitude and exposure, thus mountain habitats have greater species richness than the lowlands when compared with similar areas. This richness decreases with increasing altitude, but isolation and environmental extremes also restrict species' habitats. For instance, globally, there are 10,000 species of flowering plants in the alpine belt alone representing 4 per cent of all higher plant species even though the alpine belt covers only 3 per cent of the Earth's land area (Körner, 1995).

The highland landscapes of the MMSEA are mosaics of pastures, wetlands, forests, croplands and human settlement: a range of habitats for all life forms. In the mountains, geography promotes cultural diversity in languages, belief systems, architecture, settlement patterns and livelihood practices. People have adapted in ways that demonstrate their intimate relationship with the environment and knowledge about plants, wildlife, vegetation and ecosystems. About 720 million (12 per cent) of the global human population lives in mountain regions, and half of them are in the MMSEA region. Of the 10 per cent living above 2500m, almost all – over 70 million – live in poverty and are vulnerable to food insecurity and mountain hazards, vulnerabilities and risks (Jodha, 2005).

Socially constructed vulnerabilities result from the perception of uplands by state agencies and lowland people as sources of strategic resources for economic development such as hydropower, timber, non-timber forest products and minerals. State-owned or private enterprises have developed logging, mining and hydropower generation schemes. Dam construction for instance has directly caused loss of biodiversity and resulted in many negative social impacts including millions of people who have faced resettlement or displacement from their original homes.

Poverty has many faces and causes. Political unrest, social conflicts, unsecured citizenship, poor infrastructure and inaccessibility are some of the main causes of poverty in the Mekong region. Areas with highest levels of poverty incidence and density may have the lowest levels of inequality as in Vietnam. In northern Thailand, the highest poverty incidence and severity is associated with more remote, sparsely settled areas in montane zones largely inhabited by ethnic minority groups. Yunnan, China, and Thailand both have much lower overall levels of poverty incidence and density, but inequality in Thailand is by far the highest in the region, whereas Yunnan has relatively low levels of inequality that appear to be relatively evenly distributed (Thomas et al, 2008).

constructed vulnerabilities; the value and challenges in a rights-based approach; the risks–rights nexus towards integrating rights for living with risks; potential strategies and means to cope with, as well as rebound after, climate risks and disasters; discussion and conclusions about improving and incorporating local rights as well as addressing the adaptive capacity of upland peoples to climate change.

Climate Risks

In the mountains, people live with risk. Natural hazards such as floods, droughts, earthquakes, landslides and volcanic activity can destroy precious agricultural land and cause hunger and famine as well as disease outbreaks (Xu and Rana, 2005).

Hazardous events can either be naturally occurring, such as earthquakes, or they can be man-made, such as failure of dams (reservoir breakage). Moreover, events can be sudden, as in the case of an earthquake; or they can occur over a period of time, as in the case of soil erosion. When disaster strikes and help is needed, mountain areas are among the most difficult to access. The most vulnerable groups – the poor, ethnic people, women and children – are often hit hardest.

The risk of being adversely impacted is often a complex function of interactions among several factors including social and political economy circumstances. Lack of rights or access to resources, for example, can result in poor and marginalized groups settling near banks prone to flooding or on slopes at high risk from landslides. Moreover, lack of information is a constraint as people living in remote mountain areas (and with prevalent low literacy rates) are often less informed about climate change risks than lowland populations. Institutional capacities, opportunities for stakeholders to respond or coordinate their activities and broader political and economic systems shape how risks are governed (Lebel et al, 2006). As already mentioned, rapid economic integration and the subsequent market-orientation in the uplands has posed many socio-economic impacts for subsistence-oriented livelihoods. The risks of natural disaster add to these socio-economic implications for local livelihoods.

Risk (like vulnerability and hazard) can be defined in many different ways; we define risk as the probability of multiple harmful outcomes (impacts) of a simple hazard together with multiple causes of single outcome over a specified time period on particular individuals and social groups (Douglas and Wildavsky, 1982; Schneiderbauer and Ehrilich, 2004).

The 'risk triangle' (Crichton, 1999) has been widely used to depict the interaction between the various drivers of risk.

The impact of a hazardous event depends on the elements at risk, such as people or property (e.g. buildings) and their associated vulnerability to damage or change as a result of the event. Estimating risk is an uncertain science based on assessing probability as it involves forecasting future events whose time and location of occurrence may be largely unknown.

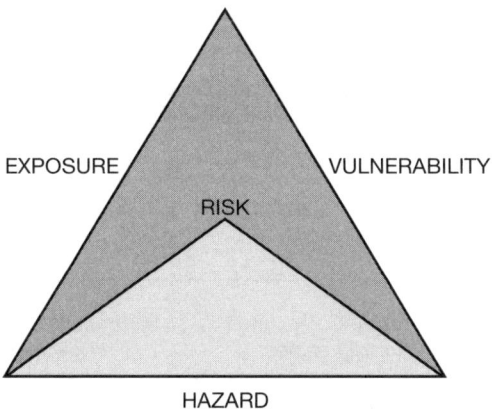

Figure 10.2 *The risk, hazard, exposure, vulnerability relationship*

Source: Crichton, 1999

Reducing the size of any one or more of the three contributing variables – the hazard, the elements exposed and/or their vulnerability – may decrease the total risk. This can be illustrated by assuming the 'dimension' of each of the three variables represents the side of a triangle, with risk represented by the area of the triangle. In Figure 10.2, the larger upper triangle portrays each of the variables as being equal, whilst in the smaller bottom halving both exposure and vulnerability has mitigated the triangle total risk. The reduction of any one of the three factors to zero would consequently eliminate the risk.

Douglas and Wildavsky (1982) argue that risk is a joint product of knowledge about the future and consent about the most desired prospects (see Figure 10.3).

When knowledge is certain and consent complete, when objectives are agreed and all alternatives (together with the probability of occurrence and consequence) are known, a programme can be written to produce the best solution. The problem is technical and the solution is one of calculation. In the next instance – knowledge is certain, but consent contested – the problem is one of disagreement about how to value consequence; the solution is either more coercion or more discussion. In the third case, complete consent is hampered by uncertain knowledge, leading to risk being defined as insufficient information and the solution is seen as more research.

Global climate change poses serious impacts in the mountain region. Although consent on the occurrence and impacts of global warming is nearly complete, knowledge is still uncertain about the locality and magnitude of impacts in the mountain regions (Xu et al, 2009). Some years or seasons might experience intense rainfall resulting in heavy floods, while other areas, years or seasons may witness less rainfall, causing severe droughts and even total failure of crops. Both flash floods as well as unexpected periods of drought can threaten loss of food and water sources. Remote mountain communities are particularly vulnerable due to poor

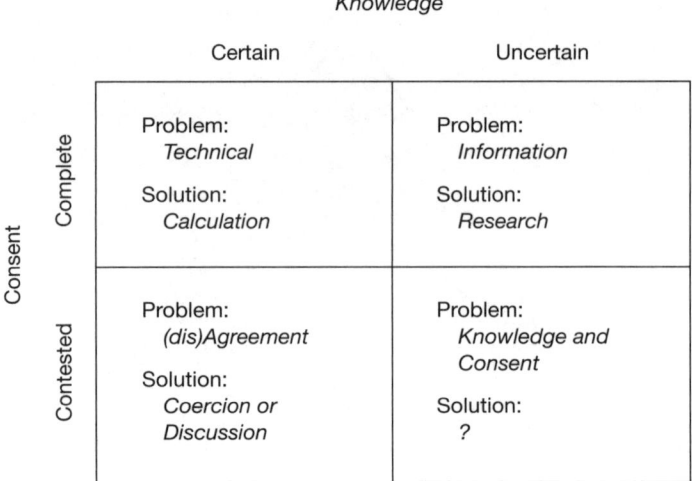

Figure 10.3 *Knowledge, risk and consent*

Source: Douglas and Wildavsky, 1982

infrastructure for protection from water-induced disasters and storage of water resources for drinking and production, as well as inadequate access to healthcare to cope with risk of disease. Climate change can result in alterations in the geographic range (latitude and altitude) and seasonality of certain infectious diseases including vector-borne infections such as malaria and dengue fever, and food-borne infections (e.g. salmonellosis), which peak in the warmer months. Zhou et al (2008) report that the waterborne parasitic disease schistosomiasis is present in the higher altitude mountain ranges in southwest China.

Climate events might trigger different impacts in the same place at different times or from place to place thus highlighting the complex and non-linear relation between climate and risk outcomes. The damages associated with climate events result more from conditions on the ground than inherently from climate variability. Climate-associated events are transformed into differentiated risk outcomes via social and political structures. Upland people who lack sufficient socio-political means are often more vulnerable to climate change risks.

Folkes et al (2005) pointed out that social learning is needed to build up experiences for coping with uncertainty and change, and concluded that learning how to sustain social-ecological systems in the context of global climate change needs an institutional and social context within which to develop and act. The risk is often shaped by society's provision of food, productive assets and social protection arrangements (Adger, 2007). A redistribution of risks is said to occur when interventions have the effect of reducing risks for one group while increasing them for another (Lebel et al, 2007b). Shifts in risk may be produced by physical changes such as embankment for flood control as well as institutional changes. New risks could arise from redistributions of hazards, exposure and/or their vulnerability. Vulnerability to hazards is to a large extent a public bad; measures to reduce vulnerability, therefore, are to a large extent, public goods (Boyce, 2000). Many measures to reduce disaster vulnerability are impure public goods, which when provided to one are provided to others, but not equally provided to all (Boyce, 2000). For example, flood-control projects provide location-specific benefits, restricted to the subset of the population who live or own assets in the protected area. Similarly, dam construction primarily benefits certain user groups while other groups of society are excluded from the benefits of these investments and even have to bear the costs such as loss of livelihoods (such as fisheries), inundation of farmlands or resettlement. Environmental impact assessments of dam construction provides insightful knowledge into the potential risks; consent is often contested because of disagreement on social consequences as well as differentiated risks, in other words different socio-economic sectors face effects in their own particular ways.

Vulnerability can be defined in terms of the capacity of individuals and social groups to prepare before a disaster occurs, to respond and cope during a disaster, to recover or rehabilitate after a disaster or adapt to any external stresses placed on their livelihoods in the risk cycle (Kelly and Adger, 2000). Since climate impacts affect disadvantaged social groups more disproportionately, the approach that we develop places the socio-economic institutions at the centre of the analysis. Local institutions influence how different social groups gain access, and are able to use assets and resources (Agrawal, 2009).

ACCELERATED RISKS AND SOCIALLY CONSTRUCTED VULNERABILITY

Climate change and accelerated risks

Recent climate modelling experiments suggest that warming would be significant in the mountains of Asia including the Tibetan Plateau, which has shown consistent trends in overall warming during the past 100 years (Yao et al, 2006; Xu et al, 2009). Various studies suggest that warming in the Tibetan Plateau has been much greater than the global average of 0.74°C over the last 100 years (IPCC, 2007a). Climate change involves, perhaps most seriously, changes in the frequency and magnitude of extreme weather events. There is widespread agreement that global warming is associated with these extreme fluctuations, particularly in combination with intensified monsoon circulations. Global El Niño/Southern Oscillation (ENSO) events have directly affected the regional annual precipitation in the Mekong uplands. Although many other factors are involved, the growing incidence and toll of related natural disasters, such as floods and droughts, are of particular concern. In the upper Mekong or the southeastern Tibetan Plateau where there is heavy precipitation and the temperature is higher than in other areas, the rising temperatures could accelerate the ablation and retreat of glaciers, potentially resulting in more frequent flash floods and landslides in upland watersheds (Xu et al, 2006).

The predicted change in climate in the Mekong region in particular, such as the changes in rainfall patterns, might be even more important as the frequency and magnitude of high intensity rainfall events increases and dry geographical areas become even drier, and dry periods last longer (Ma et al, 2009). The drought in 2010 in China for instance affected five provinces in the southwest region (Qiu, 2010) as well as the countries of Burma/Myanmar, Thailand, Lao PDR, Cambodia and Vietnam. Rainfall has been well below normal, devastating crops while forest fires are covering wider areas.

Climate change is set to bring increased uncertainty to water availability. The lack of high frequency observed data and poor exchange of information in the Mekong region has been a constraint to a comprehensive assessment of climatic impacts of extreme events; therefore disagreement exists about the problems and severity of impact outcomes. But available studies suggest changes in precipitation patterns and increase in extreme events. An increase in the frequency of high intensity rainfall is already being observed in Yunnan (Ma et al, 2009). High intensity events can lead to flash floods and landslides. Monsoons in Asia are related to large-scale climate phenomena such as El Niño and La Nina, which lead to both lesser and higher than average monsoon precipitation respectively (Dhar and Nandargi, 2003). La Nina results in a significant increase in monsoon precipitation and causes the floods while El Niño often causes droughts. A commonly perceived phenomenon in the high mountains is the shift from snow dominated to rain dominated precipitation regimes due to global warming (Barnett et al, 2005).

In the mountains, natural hazard and risks are omnipresent. Intense seasonal precipitation during monsoons often triggers hazard events at different elevations. Climate change has intensified the recurrence of devastating flash floods including landslides and debris flows in Yunnan in southwest China. Such landslides and debris flow, released by torrential rain, may cause ephemeral dams across river courses and result in the impoundment of immense volumes of water. Subsequent overtopping of, or breaking through, the earth dam will result in landslide and debris outburst floods. Since the 1980s, flash floods have caused the loss of 148 lives per year, economic losses of US$30 million per year in the 1980s, $75 million per year in the 1990s and $450 million per year in the decade of 2000 (Liu et al, forthcoming).

The Rights-based Approach

The risks or vulnerability is determined by resource availability and by the entitlement of individuals and groups to call on these resources. Individual vulnerability is determined by access to resources, education and local knowledge, and the diversity of income sources, as well as by social status of individuals or households within a community. Collective vulnerability of a nation, region or community is often determined by institutional, infrastructure and market structures (Adger, 1999).

Upland areas have greatly differing social, environmental and economic levels of development within a highly inequitable global society. Building human rights criteria into future planning for addressing the impacts of climate change would help towards better understanding of who is at risk and how we should act to protect them (ICHRP, 2008).

A rights approach enables views of a range of climate risks that can be dealt with or reduced through improving different social structure and institutions or by enhancing political opportunities including better access to information and public participation. The impacts from climate change are undermining the realization of a broad range of internationally protected human rights: rights to food, water, shelter and property; rights to health; right to a safe and healthy environment; rights associated with livelihood and culture; migration and resettlement and security in the event of conflict or resource competition (ICHRP, 2008). There is also a growing global movement to promote rights of 'access';[4] that is, peoples' access to information, decision-making and justice in environmental matters, as enshrined in the Aarhus Convention (The Access Initiative, 2005).[5]

A rights dimension ensures that steps taken to address climate change are not done at the cost of the most ecologically and politically vulnerable people. Upland areas in the Mekong region are predominantly settled by ethnic communities and in ecologically vulnerable mountain environments. The politically marginalized circumstances mean these communities are unlikely to carry much weight or influence over policy-making. A rights-based framework can help direct immediate attention to the specific impacts on Mekong upland communities both from climate change as well as policies and actions for adaptation and mitigation.

For members of ethnic and linguistic minorities, issues of vulnerability are also linked to discrimination – ethnic, linguistic or gender. An analysis sensitive to the dynamics that drive processes of exclusion and discrimination is potentially more likely to anticipate future trends and deal with vulnerabilities (Oxfam International, 2008). It is now acknowledged that social and economic vulnerability greatly increases the risk of suffering from climate change. In many remote mountain areas, those without access to public health services or development infrastructure also often lack the information or resources to make informed choices to adapt to, or avoid, vicious impacts.

As pointed out earlier in the Douglas and Wildavsky (1982) framework, knowing about risks is of minimal value unless upland peoples are able to cope with those risks, thus the need for a more rights-based approach when dealing with climate risks.

A right-based approach is more integrative, viewing climate risks as depending on both biophysical (the mountain ecosystem with its risk potential) and human actors (poverty as socially constructed). Vulnerability is viewed as having '…an external dimension, which is represented...by the "exposure" of a system to climate variations, as well as an internal dimension, which comprises its "sensitivity" and its "adaptive capacity" to these stressors' (Füssel and Klein, 2006).

An increasing challenge faced by upland communities is that mitigation scenarios that rely on global finance are pushing regional governments to embark on carbon-reduction activities such as tree plantations or forestry protection projects to earn carbon credits. These plantation schemes are often undertaken in communities where the cost of doing so is fairly low. But these projects often have human rights consequences for vulnerable people in these areas and may lead to disrupting their access to forests resources or even being resettled from their lands (Lang, 2008).

For poor communities in remote mountain areas, forest ecosystems are of particular importance to provide services including water for irrigation, pollination and pest control services that are also important to other downstream areas and sectors (United Nations Environment Programme, 2005). Forests are also the traditional livelihood safety net for local people, in addition to providing a major pillar of daily agriculture and forest-based production systems. When crops fail or other natural disasters strike, basic food and cash needs are met by collection of non-timber forest products.

The disruption of forest ecosystems damages the provision of these services and benefits as well as reducing forest-based incomes. The impact of climate change on forests can make already vulnerable communities even more vulnerable to 'natural' disasters such as forest fires, landslides and floods that will result from climate change. The loss of forest-based services and income sources due to either climate change or mitigation schemes can be viewed as the contravention of people's economic, social and cultural rights (Seymour, 2008).

Climate change in the uplands cannot be addressed only within domestic boundaries as the Mekong region's mountain ecosystems and river basins lie swathed across national borders. Large water diversion or reforestation schemes in an upland area can pose diverse and different impacts elsewhere. The rights approach helps to clarify that state obligations do not stop at their own borders: they have a responsibility to monitor the impacts on neighbouring countries of their domestic policies or where necessary regulate their private sector entities operating across borders.

However, the rights-based approach still faces a number of challenges particularly in its legal framework (ICHRP, 2008) as rights issues are often marginalized in climate change discussions. One reason is that global change generally affects categories of rights that are not known for strong enforcement mechanisms in international law: social and economic rights, the rights of migrants, the rights to a safer environment during disasters and rights protections during conflicts.

Second, human rights law places the government with the primary duty to act when rights are violated. But with climate change, responsibility for the impact on the most vulnerable people may rest not just with the nearest government but across many diffuse actors, both public and private, many of whom may be also geographically distant. Human rights law has conventionally not found it easy to impose obligations across international borders. Moreover, the distribution of costs and benefits results in different actors contesting their rights as different interest groups come into play against each other.

Mainstreaming of human rights occurred with many UN agencies, bilateral development agencies and non-governmental organizations (NGOs) taking up 'human rights-based approaches' (HREOC, 2008). In practice, adoption has been uneven as international financial institutions, multilateral development banks and private foreign investors have refused to adopt a human rights methodology (ICHRP, 2008). Applying international human rights law to these actors continues to be an uncertain exercise as they are neither states nor, so it is argued in some cases, subject to specific territorial jurisdictions.

Another weakness with international human rights law is that clear evaluation mechanisms are not available to assess development activities for their rights outcomes or to hold the principal development actors to account (Tan, 2008). Thus the development and rights relationship continues to remain complicated with their integration in practice 'at best a work in progress' (ICHRP, 2008). This also partly explains the relative neglect of rights issues in climate change discussions.

While the rights-based appraisal of climate change is receiving much emphasis, the framework relies on cost–benefit and other economic analyses. The human rights issues are drawn upon only for their normative value and to illustrate distributional justice issues. The approaches are thus neither examining specific rights violations resulting from climate change, nor considering actions to address it.

Given that rights-based issues and approaches are essential to address the vulnerabilities of people living in mountain regions, we make a tentative list of some of the critical rights issues for mountain peoples as the following:

- Inclusion and exclusion: during nation-state building processes, many ethnic groups or 'hill tribes' were considered as socio-political refugees. For example, lowland Thai officials have resisted providing citizenship papers to upland ethnic peoples for decades. In China, people are registered either in rural areas as peasants or in the urban areas as urban citizens, with totally different access to healthcare and other social welfare provisions. The boundaries of collective identity are often political
- Access and resource rights: although mountain people are resource-dependent, resources in and of themselves do not constitute security since resources are mediated through access and property rights. Access to resources is often based on social and economic relations (Blaikie et al, 1994). Access in this context can be taken to mean 'the ability to derive benefits from things by emphasizing the ability rather than the rights to benefits' (Ribot and Peluso, 2009)
- Prior information: mountain ecosystems have many micro-environments that are susceptible to change with global warming and increasing human activities.

Mountain peoples should be fully informed about potential increasing risks. They should have access to early warning systems for disaster forecasting
- Land use and mobility: mobility is one key feature of land use in the mountains and a salient feature of mountain identity. Communities have traditionally accessed different habitats during different seasons for fodder and forest products; swidden-fallow systems maintained soil fertility and controlled weeds; nomads chased their livestock in search of water and rainfall for fertile pastures. Mountain people accessed different areas of farmlands in different seasons along various altitude gradients
- Poverty and development rights: poverty (and how it is defined) is an important aspect of vulnerability because of its direct relationship with natural assets, access to resources and recovery from the impacts of extreme events. The overlap of high poverty incidence and more hazardous areas in the mountain region is due to political marginalization of mountain peoples combined with poor investments in infrastructure and failure to meet the costs for maintenance by governments
- Climate change risk governance: the process through which actors and other agencies wield power and authority to make and enforce decisions on climate change in terms of both mitigation and adaptation is significant. Risk governance is the use of political authority and exercise of control in relation to the management of climate change risks for social and economic development. The critical question is how are local voices heard and whose agenda counts in the decision-making process.

THE RISKS–RIGHTS NEXUS

It is important to understand the dynamic interactions between multiple harmful impacts of hazards, multiple risks and causes of vulnerability (the 'nexus') at multiple scales from local to regional levels. Extreme climate events trigger different natural hazards and risks across spatial and temporal scales. Climate change impacts people differently. It has already been noted that the poor are disproportionately vulnerable to loss of livelihoods and assets in the face of global change.

First, living with multiple risks, poor and marginalized groups are often more vulnerable than economically better off people. The reduction in vulnerability is most pronounced in high-income individuals, groups and classes, who have a 'willingness to pay' and will get more of the benefits of disaster-vulnerability reduction than their poorer counterparts (Boyce, 2000). The higher levels of well-being along with better infrastructure and better access to information and preparedness technology are factors in successfully mitigating risks. Environmental risks can be transformed into differentiated outcomes via social structure and rights. Jackson et al (2006) reported that the tuberculosis (TB) and poverty link in China was due to the distinct separation of rural and urban health systems. Rural people, especially women, those less covered by healthcare schemes, less educated and living farther from health facilities faced the longest delays in reaching TB services, getting diagnosis and treatment (Cheng et al, 2005).

Second, the pursuit of adaptive capacity is likely to produce side-effects and redistribution of risks (Lebel, 2007). Building large storage dams to secure water supplies for one area or group may have major implications for other water users downstream and resettlement of upstream people. The structured measures for disaster reduction might protect some groups or a majority, but increase or divert risks to others. The pursuit of adaptiveness can create winners and losers, shift risks and burdens and reinforce existing inequalities (Adger, 2007). The most vulnerable people have less negotiation power in climate change mitigation and adaptation (Paavola and Adger, 2006).

Third, different political strata experience different risks (Sen, 1981). Differences in access to critical social processes, particularly in decision-making about adaptation policies, might further marginalize mountain peoples. The diverse outcomes of climate change impacts are a consequence of different rights-based social and political-economic environments. Place-based inequality, including access to political power and social security, early warning or planning systems, translate risks into suffering and loss. Government planned adaptation efforts often have difficulty in reaching groups especially the poor and ethnic people in remote mountain regions.

Fourth, poverty is the most salient condition in the Mekong uplands that shapes risks and vulnerability. The mountain poor are least able to buffer themselves against, and rebound from, stress. They often live in the marginalized ecological zones (flood- and drought-prone environments), lack adequate social and health insurance to help them recover from losses, are least informed about global climate change risks and have little influence to demand that their governments provide protective infrastructure, temporary relief or reconstruction support.

Vulnerability analysis has two, often depicted as polar, opposites: exposure to place-based risks and hazards (risk-based approach) or social construction (lack of access). Integrative approaches have evolved over the past two decades to link these two views, with specific focus on place and region (Füssel and Klein, 2006). The risk model tends to evaluate the multiple outcomes (or 'impacts') of a single environmental event, while the livelihoods and social institutions approaches characterize the multiple causes (including climate) of single outcomes (Ribot, 1995).

We try to ask: which places (where) and type of risks (what), social groups (who) and timing of risks (when) are vulnerable, and the kinds of rights involved and affected. The question of what we need to invest in requires an understanding of the characteristics of their vulnerability and reasons (why) these places and (whose) people are at risk, so we can assess the full range of means for reducing that vulnerability. We use the case of floods to assess the risks–rights nexus in terms of pre-disaster (preparedness), during disaster (responses) and post-disaster (recovery) in case of flash floods in the mountain region (see Table 10.1). Our argument is that for these 'natural risks' such as floods, the manner and direction of development enhances or decreases the vulnerability of people to the climate risks.

Table 10.1 *The risks–rights nexus during flash flood management cycle*

Disaster cycle	Risks	Associated rights	Consequences
Pre-disaster (Preparedness)	Life and livelihood activities in flood prone areas such as landless people	Lack of access to resources, no legal rights such as land or property tenure	Socio-economic vulnerability
	High frequent exposure to flood hazards	Lack of protective infrastructure and no access to flood information	Physical vulnerability
	Increasing hazard potential due to intensified land use and change of property rights	Lack of participation in land use policy planning, no or weak alternative livelihoods	Loss of mobility, and risk-escape routing
	Redistributed risks due to dam building and river embankment	Lack of prior information and consent, and social exclusion	Increased risks and insecurity
	Increasing climate change risks with incomplete knowledge	Weak governance in climate change adaptation	Low preparedness
During disaster (Response)	Hazards including landslide, mudflows	Access to early warning and information	People perceive natural disasters as an act of God
	High exposure for particular individuals (women and children) and groups (ethnic and poor)	Poor social inclusion for the most vulnerable groups	Loss of life and property
Post-disaster (Recovery)	Dislocation and displacement	Poor access to basic needs and relief and rehabilitation programmes	Famine
	Exposure to new risks such as epidemic diseases	Lack of healthcare system	Poor human health

When flash floods occur, they pose a serious threat to upland people's lives and livelihoods (Xu et al, 2006). Vulnerable groups, including the poor, women, children and the elderly, should be given special attention with regard to flash floods and their management. But in fact, upland vulnerable people are often excluded from access to resources and institutions for disaster management.

Socially constructed vulnerability

Vulnerability can be defined as the coping capacity of individuals and social groups to respond, to recover from or adapt to any environmental stressors placed on their livelihoods and well-being (Blaikie et al, 1994). A disaster takes place when the external stressors exceed the coping capacity of the population. Such coping capacity largely depends on access to resources over time. Thus vulnerability is socially constructed (Sen, 1981). From this perspective, the risks need to take account of the analysis of individual and collective perceptions, representations of power relations and interactions of social actors and the linkage between risks and rights.

Mountains have an abundance of freshwater, forestry and minerals, as well as environmental services like hydrological benefits for downstream areas. Extractive industries value mountain ecosystems because of their mineral resources that generate revenue. Downstream extractors look for timber and non-timber products, and pharmaceutical companies gather valuable plant material. Urban consumers like mountain 'niche' products. Tourists value the natural beauty or cultural artifacts. Environmentalists strive to conserve endemic species. Dam developers take advantage of the mountain rivers' hydropower potential, but big dams leave not only large ecological footprints because of submersion, but also flood forests or displace many indigenous people from their homes and cause conflicts. However, outsiders pay scant attention to the risks mountain people face with a changing environment and their rights to access social support systems for addressing vulnerability. We base our examination of social vulnerability to global change and climate variability on an understanding of the mountain peoples, water-related risks and relationship with others.

Floods

Floods are considered as the principal hazards in the uplands, lower valleys and plains. The intensity, quantity and locations of rainfall are the key hazard elements determining the extent and magnitude of disaster along with local geology, land use and overall denudation of the basin. Where floods are carefully monitored and information released through radio and television, early warning systems can minimize impacts, for example by evacuating people from danger zones, and post-disaster relief can also reach the flooded areas and impacted people quickly. On the other hand, early warning systems are unable to cope with the magnitude and frequency of flash floods in the mountain region. Present-day scientific capability is still unable to predict and take measures to cope with flash floods. In comparison to riverine floods, water levels in flash floods rise very rapidly reaching peak flows sometimes within minutes. Flash floods can be caused or compounded by climate change and pose a serious threat to upland peoples.

Local knowledge may prove functionally inadequate in dealing with emerging climate change risks where potential shifts in monsoon rain and intensity pose risks of more frequent flash floods. In early November 2008, after the regular monsoons, sudden and intense cloudbursts triggered flash floods that swept down the mountains, resulted in large-scale landslides and debris flows, caused 28 deaths, left 40 people missing and cost nearly 1 billion Yuan of property loss in Chuxiang prefecture of Yunnan province. The lack of access to timely information such as the degree and severity of rains, possible flood events and their forecasted impacts have constrained upland peoples from adequate disaster preparedness and post-disaster rehabilitation.

Drought

In contrast to the official responsiveness to the flood problem, inadequate attention is paid to extreme drought in the mountain region. Due to the nature of mountain ecosystems, there is both extreme rainfall run-off and flash floods in the wet season and water scarcity in the dry season. The impacts of drought often hit the poor in the uplands most. The drought in 2010 in the Mekong region affected five provinces in southwest China (Qui, 2010), as well as the countries of Burma/Myanmar, Thailand, Lao PDR, Cambodia and Vietnam. The drought was so severe that river water levels were at lows not seen in the last 100 years. The drought affected 65 million people, mostly upland farmers in the Mekong region. The drought-stricken mountain provinces of southwest China have already lost 2.1 million hectares of crops while millions of livestock are at risk. The economic cost to China so far has been an estimated US$1.46 billion (see www.digitaljournal.com/article/288665, last accessed March 2010).

Waterborne diseases

Upland people have increasingly been exposed to new health risks especially from vector-based diseases like malaria (Martens et al, 1999) and schistosomiasis (Zhou et al, 2008). Malnutrition from reduction in crop yields would increase the severity of impact of these diseases. Waterborne diseases are caused by microorganisms, which are directly transmitted when contaminated freshwater is consumed by humans. Steinmann et al (2007) find out that females and individuals aged less than 10 were at a higher risk of parasitic schistosomiasis diseases; on the other hand, a higher socio-economic status was a protective factor. Global climatic change may have a significant impact on the distribution of vector-borne diseases. For instance, the infection rate for malaria is an exponential function of temperature where small increases in temperature can lead to a sharp reduction in the number of days of incubation. Regions at higher altitudes or latitudes may thus become hospitable to the vectors and may be invaded by vectors as a result of an increase in the annual temperature. Rapid development of local hydrological projects have increased the possibility of pathogen spread and establishment of new habitats (Liang et al, 2006) and have therefore increased health risks. Although there is reported evidence of the importance of hydrological connectedness and climate change associated with the transport of waterborne diseases, increased socio-economic connectivity through increasing mobility of humans and livestock also plays an important role in spreading infectious pathogens (Liang et al, 2006; 2007).

TOWARDS INTEGRATING RIGHTS FOR ADDRESSING RISKS

Certain risks cannot be eliminated by human ingenuity alone, but human beings can reduce the possibility of occurrence and people's capacity to address the potential consequences by improving the rights of vulnerable individuals and groups and enabling them to better deal with those risks. Floods depend upon precipitation, but also upon banking of rivers, drainage, bio-stabilization of slopes and proper retaining walls and terracing of fields. If implemented appropriately, through active participation of local citizens and community-based disaster risk management, such measures can also help lessen damage from landslides, rockfalls and mudflows (Xu et al, 2006). Local practices of using mountain plants, shrubs and trees capable of anchoring unstable soils has led to the more modern practices of bio-technical stabilization of slopes to control landslides and soil erosion. Local knowledge systems including folklore and customs have ways to cope with these hazards, consisting of rotating periods of guardianship in which certain groups are charged with protecting or warning the community. Good governance in climate change adaptation and mitigation should include the participation of vulnerable groups and inclusion of their knowledge systems. The traditional wall-building techniques and patterns that farmers and herders in the highlands used to hold back possible landslides and rockfalls can be found today in the skills of engineering geologists who study mountain risk engineering (Xu and Rana, 2005).

Mountain trails are cared for by communities who understand the hazards inherent in shepherding flocks and herds along the high precipices. These same trails, used by traditional livelihood processes that had adequate responses to impacts, become hazardous when used by modern tourism. Mountain people today not only have to cope with the disaster risks that they expect, and for which they have traditional coping mechanisms, but in an era of economic change, mountain people are also leaving their homes to seek more opportunities. Living with risks in the mountain regions can be as much from the social and economic structures as from the physical environment.

Mobility of livelihoods and social networks are examples of adaptation. Seasonal migration occurs when risks are greatest. Transhumance, the ancient practice of moving herds from summer to winter pastures, ensured that pastures were not overgrazed, thereby preventing erosion. The drought in 2010 in the

Mekong region has triggered increasing seasonal migration of rural people looking for off-farm opportunities, such as hired labour, in the cities.

Remote sensing and global positioning systems assist in mapping landslides, monitoring floods and droughts, calculating and forecasting the probability of extreme climatic events and assessing groundwater potential. Governments and non-government organizations can play bigger roles in institutional arrangement, capacity-building and knowledge transfers for risk management. The private sector and civil society are important partners in all phases of the disaster management cycle: prevention/mitigation, preparedness, response and recovery. These assist, but do not replace, the watchfulness of mountain communities, their knowledge of environmental signs of impending disaster risks or their institutions for coping with climate change induced risks. Nor do modern warning systems replace entirely their traditional means of communication and learning. Traditional groups and their leaders have become forest and water user groups, and the risk management groups of today as local institutions adapt to changing ecological circumstances and challenges (see Chapter 4, this volume). In the mountains, there is a convergence of traditional and modern knowledge with supportive local institutions that strives to enable communities to constantly adapt and live safely in the highest places on Earth.

Discussion and Conclusions

Mountain habitats have always been subjected to both positive and negative impacts of global change, with consequences for human well-being. Valued not only for the goods and services received downstream, but also for their economic livelihood, cultural and spiritual values, mountains capture the imagination (often the mountain environment more than the people). Yet mountain dwellers have adapted and are adapting to habitats, hazards and risks in the some of the most difficult and fragile terrains on earth. However, their capacity to respond to global drivers at different scales of environmental change is limited. National policies and strategies that integrate their rights and access to resources into efforts to adapt and deal with climate change in the uplands are needed.

Mountain peoples need to be fully informed about the drivers and impacts of global climate change, particularly the consequences and outcomes that affect their livelihoods. Importantly, upland peoples need to be better able to influence the development decisions, in particular those taken in the name of climate risks management, which might affect them. Equally necessary is the creation of mechanisms for recourse from socially constructed risk, for example access to special emergency funds for risk management or adaptation.

Support for risk management is needed from the lowland and urban decision-makers. Local resources must be made accessible to mountain people in maintaining resilient livelihoods. Livelihood strategies and opportunities locally for the mountain poor must take pressure away from limited land resources. The impacts in the uplands of large-scale dam and upland infrastructure, and the extraction of timber and mineral resources, need to be critically examined in terms of sustainability. Community-based disaster risk management needs to be reassessed where these risks place unfair burdens on the poor. For those living in remote upland areas and away from urban prosperity, the risks from natural disasters or from adverse changes to upland ecosystems have often not reduced their risks. This is because newer involuntary risks are redistributed towards them while wealthier urban inhabitants can afford to cope by using infrastructure, insurance and compensation. Community-based disaster risk management needs to be further strengthened with supportive access to information and technology.

Through integrated research and careful development interventions, environmental impact assessments are being used more frequently before undertaking large infrastructural or resource exploitation in mountains areas. Environmentally friendly enterprises such as beekeeping with indigenous bees, the production of silkworms and silk, off-season vegetable and fruit farming and medicinal herbs and spices are well-suited to mountain areas and can be handled without heavy infrastructure that can limit climate disaster risks.

In this chapter, we have described the challenges and contradictions of living in a mountain habitat and argued for an approach that links mountain risks and incorporating the rights of uplands peoples as a way of addressing climate-related impacts. Mountain peoples, their rights and their traditional methods of addressing risks need to be the focus of concerted and genuine development efforts to alleviate vulnerabilities inherent in the mountain environment and to help them cope with the future impacts of climate change.

Acknowledgements

This work was funded by MISEREOR, an overseas development agency of the Catholic Church in Germany 'against hunger and disease in the world', and the Mekong Program on Water, Environment and Resilience (M-POWER), supported by the CGIAR Challenge Program on Water and Food, Echel-Eau and the International Fund for Agricultural Development (IFAD).

Notes

1 See www2.ohchr.org/english/law/cescr.htm (last accessed 22 July 2009).
2 The Mekong uplands, lying within the larger context of Montane Mainland Southeast Asia (MMSEA), as defined in this chapter, is a large, eco-region comprising about half of the land area of Cambodia, Laos, Burma/Myanmar, Thailand, Vietnam and China's Yunnan Province. The headwaters of the Yangtze, Salween, Irrawaddy, Mekong, Red,

Chao Phraya and Pearl Rivers are located within the MMSEA region that drain an area of nearly 4 million km^2 and have impacts on the lives of more than 696 million people (Thomas et al, 2008). The MMSEA region can be further divided into an alpine zone (above 3000masl), a high mountain zone (1000–3000masl) and a low mountain zone (300–1000masl). The term 'uplands' is used here to describe areas in the montane and alpine zones (Thomas et al, 2008). Our definition centres on areas that are 300–3000 metres above sea level (masl) in elevation, and located within and across several river basins. The alpine zone, which is dominated by the high altitude Tibetan Plateau, is referred to as the 'Water Tower of Asia' (Xu et al, 2008), while the montane zone has been called the 'Roof of Southeast Asia' (Thomas et al, 2008).
3   Climate change will have implications for the enjoyment of human rights. The United Nations Human Rights Council recognized this in its 'Human rights and climate change' (28 March 2008, available at www2.ohchr.org/english/issues/climatechange/docs/Resolution_7_23.pdf, accessed 22 July 2009), expressing concern that climate change 'poses an immediate and far-reaching threat to people and communities around the world' and requesting the Office of the United Nations High Commissioner to prepare a study on the relationship between climate change and human rights (www2.ohchr.org/english/issues/climatechange/index.htm, accessed 14 June 2010).
4   For instance, The Access Initiative (TAI) is a global network of non-governmental organizations working to ensure that people have the right and ability to influence decisions about the natural resources that sustain their communities. Working in their respective countries, TAI partners form national coalitions assess the performance of their governments to provide the public with access to information about government decisions, public participation in decision-making and access to justice when their rights to information, participation and a clean environment are violated. The right to obtain government information, right to participate in government decision-making and the right to seek justice are a bundle of valuable rights that is called 'access rights' (see www.accessinitiative.org/about, accessed 22 July 2009).
5   The United Nations Economic Commission for Europe (UNECE) Convention on Access to Information, Public Participation in Decision-making and Access to Justice in Environmental Matters, usually known as the Aarhus Convention, was signed on 25 June 1998 in the Danish city of Aarhus. The Aarhus Convention grants the public rights regarding access to information, public participation and access to justice, in governmental decision-making processes on matters concerning the local, national and transboundary environment. The Convention has a Compliance Review Mechanism, unique in international environmental law, that allows members of the public to communicate concerns about a party's compliance directly to a committee of international legal experts empowered to examine the merits of the case (www.unece.org/env/pp/, accessed 22 November 2010).

References

Access Initiative (2005) Citizen Voices in Water Sector Governance: The role of transparency, participation and government accountability, www.accessinitiative.org

Adger, W.N. (1999) 'Social vulnerability to climate change and extremes in coastal Vietnam', *World Development*, vol 27, no 2, pp249–269

Adger, W.N. (2007) 'Vulnerability', *Global Environmental Change*, vol 16, pp268–281
Agrawal, A. (2009) 'The role of local institutions in adaptation to climate change', in R. Mearns and A. Norton (eds) *Social Dimensions of Climate Change: Equity and Vulnerability in a Warming World*, The World Bank, Washington DC
Barnett, T.P., Adam, J.C. and Lettenmaier, D.P. (2005) 'Potential impacts of a warming climate on water availability in snow-dominated region', *Nature*, vol 438, no 17, pp303–309
Bates, B.C., Kundzewicz, Z.W., Wu, S. and Palutikof, J.P. (eds) (2008) 'Climate change and water', technical paper of the Intergovernmental Panel on Climate Change, IPCC Secretariat, Geneva, Switzerland, p210
Blaikie, P., Cannon, T., Davis, I. and Wisner, B. (1994) *At Risk: Natural Hazards, Peoples Vulnerability and Disasters*, Routledge, London
Boyce, J.K. (2000) 'Let them eat risk? Wealth, rights and disaster vulnerability', *Disasters*, vol 24, no 3, pp254–261
Cheng, G., Tolhurst, R., Li, R., Meng, Q. and Tang, S. (2005) 'Factors affecting delays in the tuberculosis diagnosis in rural China: A case study in four counties in Shandong Province', *Transactions of the Royal Society of Tropical Medicine and Hygiene*, vol 99, pp355–362
Crichton, D. (1999) 'The risk triangle', in J. Ingleton (ed.) *Natural Disaster Management*, Tudor Rose, London, pp102–103
Dhar, O.N. and Nandargi, S. (2003) 'Hydrometeorological aspects of floods in India', *National Hazards*, vol 28, pp1–33
Douglas, M. and Wildavsky, A. (1982) *Risk and Culture: An Essay on the Selection of Technological and Environmental Dangers*, University of California Press, Berkeley
Folkes, C., Hahn, T., Olsson, P. and Norberg, J. (2005) 'Adaptive governance of social-ecological systems', *Annual Review of Environment and Resources*, vol 30, pp8.1–8.33
Füssel, Hans-Martin and Klein, Richard J.T. (2006) 'Climate change vulnerability assessments, an evolution of conceptual thinking', *Climate Change*, vol 75, pp301–329
HREOC (2008) *Human Rights and Climate Change*, Human Rights and Equal Opportunity Commission (HREOC), Australian Human Rights Commission, Sydney, Australia
ICHRP (2008) *Climate Change and Human Rights: A Rough Guide*, International Council on Human Rights Policy, Geneva
IPCC (2007a) 'Climate Change (2007): Impacts, Adaptation and Vulnerability', contribution of Working Group II to the Fourth Assessment Report of the IPCC, Cambridge University Press, Cambridge, UK
IPCC (2007b) 'Climate Change (2007): The Physical Science Basis', contribution of Working Group I to the Fourth Assessment Report of the IPCC, Cambridge University Press, Cambridge, UK
Jackson, S., Sleigh, A.C., Wang, G.J. and Liu, X.L. (2006) 'Poverty and the economic effects of TB in rural China', *International Journal of Tuberculosis and Lung Disease*, vol 10, no 10, pp1104–1110
Jodha, N. (2005) 'Adaptation strategies against growing environmental and social vulnerabilities in mountain areas', *Himalayan Journal of Sciences*, vol 3, no 5, pp33–42
Kates, R.W. (2000) 'Cautionary tales: Adaptation and the global poor', *Climate Change*, vol 45, pp5–17

Kelly, P.M. and Adger, W.N. (2000) 'Theory and practice in assessing vulnerability to climate change and facilitating adaptation', *Climatic Change*, vol 47, pp325–352

Körner, C. (1995) 'Alpine plant diversity: A global survey and functional interpretations', in F.S. Chapin III and C. Körner (eds) 'Arctic and Alpine Biodiversity: Patterns, Causes and Ecosystem Consequences', *Ecological Studies*, vol 113, pp45–62

Lang, C. (2008) 'Taking the land, impoverishing the people: The pulp industry in the Mekong Region', *Watershed*, vol 12, no 3, November, pp92–101

Lebel, L. (2007) 'Adapting to climate change', *Global Asia*, vol 2, pp15–21

Lebel, L., Nikitina, E., Kotov, V. and Manuta, J. (2006) 'Assessing institutionalized capacities and practices to reduce the risks of flood disasters', in J. Birkmann (ed.) *Measuring Vulnerability to Natural Hazards: Towards Disaster Resilient Societies*, United Nations University Press, Tokyo, pp359–379

Lebel, L., Sinh, B., Garden, P., Hien, B., Subsin, N., Tuan, L. and Vinh, N. (2007b) 'Risk reduction or redistribution? Flood management in the Mekong region', USER Working Paper WP-2007-10, Unit for Social and Environmental Research, Chiang Mai University, Chiang Mai

Liang, S., Yang, C., Zhong, B and Qiu, D. (2006) 'Re-emerging schistosomiasis in hilly and mountainous areas of Sichuan', *China Bulletin World Health Organisation*, vol 84, pp139–144

Liang, S., Seto, E.Y., Remais, J.V., Zhong, B., Yang, C., Hubbard, A., Davis, G.M., Gu, X., Qiu, D. and Spear, R.C. (2007) 'Environmental effects on parasitic disease transmission exemplified by schistosomiasis in western China', Proceedings of the National Academy of Sciences, USA, vol 104, pp7110–7115

Liu, H., Zhou, Y., Zhu, H. and Zeng, Z. (forthcoming) 'Impacts of climate change on mountain disasters and its adaptive measures', in Y. Zhou, X.X. Lu, J.C. Xu, H. Zhang and T. Jiang (eds) *Systematic Assessment of Climate Change Impact in Yunnan Province*, China Meteorological Press (in Chinese), Beijing

Ma, X., Xu, J.C., Luo, Y., Aggarwal, S.P. and Li, J.T. (2009) 'Response of hydrological processes to land-cover and climate changes in Kejie watershed, SW China', *Hydrological Processes*, doi10.1002/hyp.7233

Martens, P., Kovats, R.S., Nijhof, S., de Vries, P., Livermore, M.T.J., Bradley, D.J., Cox, J. and McMichael, A.J. (1999) 'Climate change and future populations at risk of malaria', *Global Environmental Change*, vol 9, ppS89–S107

Molle, F., Lebel, L. and Foran, T. (2009) 'Contested Mekong waterscapes: Where to next?' in F. Molle, T. Foran and M,. Käkönen (eds) *Contested Waterscapes in the Mekong Region: Hydropower, Livelihoods and Governance*, Earthscan, London

O'Brien, K., Eriksen, S., Nygaard, L.P. and Schjolden, A. (2007) 'Why different interpretations of vulnerability matter in climate change discourses', *Climate Policy*, vol 7, pp73–88

Oxfam International (2008) 'Climate Wrongs and Human Rights: Putting people at the heart of climate-change policy', Oxfam International, London, UK

Paavola, J. and Adger, W.N. (2006) 'Fair adaptation to climate change', *Ecological Economics*, vol 56, no 4, pp594–609

Qui, J. (2010) 'China drought highlights future climate threats', *Nature*, vol 465, no 13, pp142–143

Ribot, J.C. (1995) 'The causal structure of vulnerability: Its application to climate impact analysis', *GeoJournal*, vol 35, no 2, pp119–122

Ribot, J.C. (2009) 'Vulnerability does not just fall from the sky: Toward multi-scale pro-poor climate policy', in R. Mearns and A. Norton (eds) *Social Dimensions of Climate Change: Equity and Vulnerability in a Warming World*, The World Bank, Washington, DC

Ribot, J.C. and Peluso, N.L. (2009) 'A theory of access', *Rural Sociology*, vol 68, no 2, pp153–181

Schneiderbauer, S. and Ehrlich, D. (2004) 'Risk, hazard and people's vulnerability to natural hazards: A review of definitions, concepts and data', Office for Official Publication of the European Communities, Luxembourg

Sen, A. (1981) *Poverty and Famine: An Essay on Entitlement and Deprivation*, Oxford University Press, Oxford

Seymour, F. (2008) *Forests, Climate Change, and Human Rights: Managing Risk and Trade-offs*, Center for International Forestry Research, Bogor

Steinmann, P., Zhou, X.N., Matthys, B., Li, Y.L., Li, H.J., Chen, S.R., Yang, Z., Fan, W., Jia, T.W., Vounatsou, P. and Utzinger, J. (2007) 'Spatial risk of profiling of Schitosoma japonicum in Eryuan county, Yunnan province, China', *Geospatial Health*, vol 2, no 1, pp59–73

Tan, C. (2008) 'Mandating rights and limiting mission creep: Holding the World Bank and the International Monetary Fund accountable for human rights violations', *Human Rights and International Legal Discourse*, vol 2, no 1, pp79–116

Thomas, D.S.G. and Twyman, C. (2005) 'Equity and justice in climate change adaptation amongst natural-resource-dependent societies', *Global Environmental Change*, vol 15, pp115–124

Thomas, D.E., Ekasingh, B., Ekasingh, M., Lebel, L., Minh Ha, H., Ediger, L., Thongmanivong, S., Xu, J., Sangchyoswat, C. and Nyberg, Y. (2008) 'Comparative assessment of resource and market access of the poor in upland zones of the Greater Mekong Region', ICRAF, Chiang Mai

UNHCHR (2009) 'Annual report of the UNHCHR and reports of the Office of the High Commissioner and the Secretary-General: Report of the Office of the UNHCHR on the relationship between climate change and human rights', UN General Assembly A/HRC/10/61, 15 January 2009

United Nations Environment Programme (2005) *Millennium Ecosystem Assessment*, Island Press, Washington, DC

Xu, J.C., Eriksson M., Ferdinand, J. and Merz, J. (2006) 'Managing flash floods and sustainable development in the Himalayas', report of an international workshop held in Lhasa, PRC, 23–28 October 2005, International Centre for Integrated Mountain Development (ICIMOD), Kathmandu

Xu, J.C. and Rana, G.M. (2005) 'Living in the Mountains', in T. Jeggle (ed.) *Know Risk*, UN Interagency secretariat of the International Strategy for Disaster Reduction, Geneva, pp196–199

Xu, J.C. and Melick, D. (2007) 'Rethinking the effectiveness of public protected areas in southwestern China', *Conservation Biology*, vol 21, no 2, pp318–328

Xu, X., Lu, C., Shi, X. and Gao, S. (2008) 'World water tower: An atmospheric perspective', *Geophysical Research Letters*, 35, L20815, doi: 10.1029/2008GL035867

Xu, J.C., Grumbine, R., Shrestha, A., Eriksson, M., Yang, X., Wang, Y. and Wilkes, A. (2009) 'The melting Himalayas: Cascading effects of climate change on water resources, biodiversity, and human livelihoods in the Greater Himalayas', *Conservation Biology*, vol 23, pp520–530

Yao, T.D., Guo, X.J., Lonnie, T., Duan, K.Q., Wang, N.L., Pu, J.C., Xu, B.Q., Yang, X.X. and Sun, W.Z. (2006) 'd18O record and temperature change over the past 100 years in ice cores on the Tibetan Plateau', *Science in China: Series D Earth Science*, vol 49, no 1, pp1–9

Zhou, X.N., Yang, G.J., Yang, K., Wang, X.H., Hong, Q.B., Sun, L.P., John, B.M., Thomas K.K., Robert, B.N. and Jurg, U. (2008) 'Potential impact of climate change on schistosomiasis transmission in China', *The American Society of Tropical Medicine and Hygiene*, vol 78, no 2, pp188–194

Part V

Conclusion

11

Ensuring Justice in Water Governance in the Mekong Region

Bernadette P. Resurreccion, Nga Dao, Kate Lazarus and Nathan Badenoch

People's rights to access water resources and secure their livelihoods in the Mekong region are being undermined by inappropriate development strategies and discourses that privilege economic growth and development over wider social welfare needs. Development of water resources infrastructure is being pursued for hydropower, irrigation and industrial uses that has not adequately taken into account impacts on the environment or small-scale water users, in particular, fishers, riparian and upland farmers. As a consequence, these social groups are likely to become even more vulnerable despite living in countries whose economies are growing and where aggregate well-being is otherwise improving. This socially unjust and disproportionate allocation of costs, risks and burdens arise for many reasons, and the chapters in this book underline several.

First, if left unchecked, development reinforces existing social differences, identities and hierarchies. These translate into serious asymmetries in resource access rights, benefits and representation based on intersecting gender, ethnic and class locations and identities. For instance, Lebel et al (see Chapter 6, this volume) demonstrate that benefits from fishery livelihoods are relatively weaker and unstable for poor women who engage in them, as their livelihoods are largely contingent on combined market and cultural factors. The Hmong and Karen people in Thailand's northern uplands (see Chapter 4, this volume and Chapter 5, this volume) are cases that highlight ethnic community knowledge and forest practices as a marker for specific water management practices that stand at odds with current conservation policies, and as such, are dismissed as inappropriate.

Second, participatory rhetoric invoked by planners may mask practices of social exclusion, thus further silencing those whose voices are conventionally muted. In an era where participatory decision-making processes are widely considered as a corrective to top-down governance, the rhetoric on participation can misleadingly hide otherwise business-as-usual exclusionary practices in water resources management (see Chapter 2, this volume). In other contexts, as shown in Chapter 3 on the Son La Dam in Vietnam, decision-making processes still elude specific stakeholders; thus wider engagement is weak and narrow economic interests define water management and resettlement agendas. In the Son La Dam, the affected communities were not involved in decision-making from the outset, resulting in problem-ridden and difficult conditions of resource access and livelihoods following their resettlement.

Third, national planning and development that narrowly privilege economic growth and accumulation may potentially undermine social welfare issues, chiefly that of livelihood and food security. In particular, hydropower development may have serious implications on fishery resources as a source of nutrition and livelihoods. This raises critical questions about unequal and lopsided development: whose rights to food are being undermined, and for whose benefit (see Chapters 7 and 3, this volume)?

Fourth, growing competition over water resources may adversely impact particular stakeholders, their environments, access to water resources and livelihood security especially in rapidly changing contexts of climate and development. In the meantime, increasing private sector-led industrialization in peri-urban areas has led to growing competition over the use of water that poses serious challenges for the sustainability of common water resources, agricultural production and food security of poor populations (see Chapter 8, this volume). Similarly, at a transboundary scale, inter-government negotiations over water allocation agreements and disaster management protocols have become increasingly uneasy as climate changes affect the region and increase uncertainty and insecurity (see Chapters 9 and 10, this volume). Rapidly growing Mekong national economies coupled with increasing population densities in the lowland areas will put further pressure on the highlands for lowlanders' secure access to water, as this access is increasingly at risk due to climate change-related droughts and longer dry spells, and thus may potentially exacerbate already existing upstream and downstream dynamics and tensions.

The occurrences of unfair allocation of risks and benefits in the use of water resources offered above underscore the need for greater justice and for enabling water rights of disadvantaged groups. In summary, they point to the need for recognizing and redressing social difference and inequality in resource use and access; meaningful and inclusive participation in decision-making; reworking national and regional development priorities and policy decisions to equally privilege growth, social inclusion and welfare; and ensuring equity in the access to water resources across all users and geographical divides especially in view of the uncertainties of climate changes. Against the unjust occurrences in water resources access and management outlined in earlier paragraphs, there is equally compelling need to reassess the parameters of justice premised on former principles observed and practiced by nation states, including those in the Mekong region. It may be reasonable to rethink these principles, and whether they still apply to situations of injustices experienced by people in the Mekong region in the domain of water resources access and management. For this, the work of Nancy Fraser (1996; 2005; 2008) has been particularly instructive, and from which a few key points will be highlighted in the following discussion below.

THE POLITICS OF FRAMING JUSTICE

Some chapters in this volume show that the risks posed by large-scale infrastructure such as hydropower dams are being, or will be, largely borne by poor and disadvantaged groups (see Chapter 3 and 7, this volume), even as state and non-state players reap or envisage huge profits. Some hydropower developers have, however, attempted to mitigate or cushion some of these social costs and risks on the part of the disadvantaged groups, but have achieved uneven or even patchy success (Lawrence, 2009; Middleton, et al, 2009; Foran, et al, 2010). These unequal costs, risks and benefits refer to a crisis in 'distributive justice', which has been conventionally premised on economic inequality, thus commanding recognition and efforts for 'redistribution' or *redistributive* justice (Fraser, 1996; 2008).

Fraser (2008) tells us that there are limits to the framing of redistribution based on economic inequality alone. For instance, Hue and Sajor (see Chapter 8, this volume) draw our attention to the health and environmental costs of state-endorsed industrialization through the use of artisanal water for rapidly emerging village-based craft industries, but which offer more economic benefits to those villagers who engage in them. Moreover, the chapters (see Chapters 4 and 5, this volume) on ethnic minority communities in northern Thailand show that these groups are constrained by state laws, more privatized regimes of water access, and official development strategies as they cope with water scarcity in the uplands and exercise their own indigenous watershed management rights and strategies. Lebel et al (see Chapter 6, this volume) underscore how gender – as it intersects with class – is a primary way with which to understand the uneasy transition from capture fisheries to aquaculture, and whether and how these have translated into equitable benefits for poor women and men. These cases underscore that former framings of distributive justice based on notions of social equity do not fully 'fit' with struggles for the recognition of social and cultural differences, as well as identity, between and among particular groups of people and their access to and use of water resources. Justice thus encompasses a socio-cultural dimension whose corresponding injustice is misrecognition of social differences based on ethnicity and gender and other social categories (Fraser, 1996; 2008). Socio-cultural recognition thus becomes a call for justice.

Alongside the need for redistribution and recognition is the related need for representation. Floch and Blake and Tran (see Chapters 2 and 3, this volume) all underscore the abuse and use of public participation in hydropower development contexts, and the politics around 'who can rightfully speak?' From the lens of the need for greater consultative and inclusive democracy, the associated injustice is often that of misrepresentation or 'political voicelessness' (Fraser, 2008). People can be impeded from full participation by rules that deny them equal voice in public deliberations and decision-making (see Chapter 3, this volume). Or in Floch and Blake (see Chapter 2, this volume), participation is partial and tokenistic, and thus suggests that participation is far from straightforward and can be subject to power dynamics. Thus, particular stakeholders suffer from political injustice, misrepresentation or complete denial of an equal voice in decision-making.

As demonstrated in various chapters of this book, 'water injustice' in contemporary times arises out of the failure of the following: *redistribution* of the risks, costs and benefits of water management in economic and livelihood terms; *recognition* of differences in socio-cultural terms that result in gender, ethnic, and class hierarchies and inequities in the benefits and uses of water resources; and *representation* in political terms that edge out and marginalize particular water resources stakeholders who are already socially differentiated and marginalised.

Many of the findings of the chapters in this volume also support the notion that in today's globalized world, the former Westphalian notion of justice where 'the common framework that determined patterns of advantage and disadvantage was the constitutional order of the modern territorial state' (Fraser, 2005, p14) is quickly losing plausibility. Water use and management increasingly transcends territorial borders. Decisions taken in one territorial state – such as perhaps in electricity planning in Thailand – often impact on the lives of people in another state. Thus, ethnic communities have been resettled due to hydropower dam development in the Nakai Plateau in Laos for the purpose of creating new networks of energy trade between Thailand and Laos. Framing and meting out of justice also have territorial and spatial dimensions, where the exercise of narrowing the territory of justice to the nation state is increasingly no longer workable. Former framings of justice have therefore been proven to be contestable, and as Floch and Blake (see Chapter 2, this volume) posit, the inherently political nature of water resources planning and control underpins decisions made in the real world. Yet paradoxically, current approaches serve to 'de-politicize' water management and governance and to mould these into more economistic and efficient operational water governance 'systems' (Roth et al, 2005). This chimes with other studies on poverty that argue for a reorientation of planning for poverty alleviation towards an understanding of power and social relations as undermining and influencing people's efforts to get rid of chronic poverty (Green, 2006; Harris, 2006). The section that follows will suggest possible pathways of more fully understanding and thus reforming existing water management institutions through an understanding of power and social relations and their mechanisms, and which necessarily involve issues of redistribution, recognition and representation outlined above.

IMPROVING WATER GOVERNANCE: HOW CAN CHANGE HAPPEN?

A number of water management practices and priorities have led to socially unjust and inequitable outcomes, many of which have been presented in this collection of papers. They allude to the need for improved water governance, and we put forward a few ideas that could enable this. First, past interventions focused mainly on creating more efficient allocation systems of water supply access and services delivery, thereby ignoring the wider social, cultural and political context that in the first place lead to the forms and outcomes of these interventions and practices (Roth et al, 2005). Second, over the years, there seems to be increased optimism that community participation is sufficient to ensure that the goals of social equity and justice in the access and use of water resources are met (Ostrom, 1990; Agarwal, 2000). While well-intentioned, these participatory approaches do not fully recognize that there are constraints to people's abilities to take effective action, or that they are socially embedded in entrenched patterns of social and political inequality, which in turn influence and constrain the way people behave and take action (Granovetter, 1992; Cleaver, 1999). Thus, it is vital to understand how participation happens and what shapes or reshapes people's participation in various water governance contexts and activities.

In similar vein, Franks and Cleaver (2007) underscore the need to first understand the practices of water governance and, as Flyvbjerg (2002) argues, focusing on 'what is actually done' offers better prospects than focusing on 'what should be done'. A better understanding of water governance practices is useful for stakeholders to identify opportunities for adopting and actualizing more redistributive and representative principles of social justice in water governance wherever relevant. It also will help in reducing the burden on those that are more palpably socially exclusive and risky to social welfare. A framework that could lay bare the processes in water governance was developed by Franks and Cleaver (2007, p163), where they point to a set of interconnecting elements: *resources* are the material (natural environment, human labour and skills, capital) and non-material properties of social systems (systems of rights and entitlements, structures) upon which different people draw when they access water. Resources encompass general relationships of power, structures of inequality and 'rules' of social life and resource allocation. This also may take the form of policy, legislation, institutions, rights, knowledge, as well as material forms such as finance, technology, natural capital and capacities. *Mechanisms* include context-specific formal and informal institutions that have been organized to access water resources, often involving negotiation, and which are also dynamic and change in response to changing conditions (sometimes these mechanisms take the form of water management committees and rules, formal and informal user groups, tariffs, maintenance funds, local and customary land and water rights, technologies). *Outcomes* refer to accessed water resources and impacts on access, livelihoods, well-being, social capital and political voice. Outcomes may also include environmental outcomes such as patterns of hydrological flows, water quantity, quality, depletion and degradation. People interact as agents at all points of the framework, shaping and being shaped by the resources, mechanisms and outcomes through a range of socially differentiated and power-laden processes.

'Resources' in the form of policies and economic trends at the global and national levels set and organize the context for water governance arrangements (Ahlers and Zwarteveen, 2009). Mekong governments, especially with the recent influence of large private corporate actors, have been favouring policies to drive economic growth through promoting large hydropower and irrigation infrastructure, and other water resources development projects, placing priority on commercial viability over social welfare and environmental sustainability (Middleton et al, 2009). Policies that promote commercialization and privatization of water resources, for example, could translate into tariffs and price mechanisms, especially for irrigation systems or for the marketing of hydroelectricity from one country to another in the Mekong region. Assumptions underlying such 'neoliberal' policies include ideas that economic rationality and efficiency are the most suitable development paradigms for water management, thus sidestepping important social differentiating factors that shape outcomes of unequal access and risks such as gender, class and ethnicity (Ahlers, 2005). As a result, water is viewed purely as an economic commodity while neglecting its other uses and benefits for livelihoods such as capture fisheries (Cleaver and Hamada, 2010; see Chapter 7, this volume). An overall policy environment that chiefly favours the trade of hydroelectricity over social welfare and livelihood security goals has meant that even efforts at community-based participatory mechanisms such as local fisheries groups asserting the fishery rights of people can only be tenuous.

Scale also matters in constituting a just regime of water governance especially in the context of the Mekong region where water systems have interconnecting flows that go beyond political territorial boundaries. Challenges to water decisions and management, for instance, occur at multiple levels and scales from upstream to downstream, region to states, states to communities and within communities. For example, upper tributary watersheds provide a range of goods and services not just to the upland people who live there but also to others downstream as well as to society at broader spatial and multiple temporal scales (Lebel et al, 2005; Forsyth et al, 2008; also see Chapters 5 and 10, this volume).

Franks and Cleaver (2007) further argue that the current emphasis on increased participation and representation, as a *mechanism* for people's water access and asserting water rights is insufficient. The social conditions (*resources and structures*) within which people's participation and actions are embedded in the first place need to be interrogated. For instance, programme interventions that mobilize the participation of poor women under prevailing conditions of gender inequality, where women in the first place shoulder disproportionate and multiple workloads of care, production and social reproduction, will pose major constraints on their abilities to meaningfully influence decision-making on a new irrigation facility or on the construction of a reservoir beside a hydropower generating station. Or, a decision to put a new tariff policy in place for water supply and distribution posits an altogether class- and gender-neutral context while in reality tariffs demand the ability to command cash, and which the poor and possibly women may have little of. These structural conditions and resource deprivations are inherently profound forms of exclusion in water governance, and depict instances where more socially just perspectives and actions are urgently necessary and can be useful. Deliberately crafted yet negotiated redistributive and welfare-oriented policies in water governance that recognize difference and enable inclusive representations – shaped in large part by paradigm shifts in favour of social equity and welfare – are thus potentially useful in realizing more just water governance regimes.

As currently advocated, participation may potentially be a tipping point in favour of fairer water governance regimes. However, as mentioned earlier, heavily relying on disadvantaged people to assert their rights to water in participatory local institutions such as river basin committees or community fisheries by itself may not fully mitigate the adverse social and environmental effects of policies that explicitly favour large-scale water projects over social welfare agendas. Intervention strategies that constitute new and alternative 'mechanisms' of participatory access and claim-making are inevitably embedded in social and development contexts (both at local and wider scales). Relations of difference and power in turn, shape these contexts, where women and ethnic groups, for instance, may in reality be constrained to act. At a higher scale, the challenge then will be to reconfigure existing relations of power behind policy and decision-making. All these point to water governance being a profoundly political exercise. Mechanisms of participation and reframing of rights and policies must go hand in hand.

Ensuring justice in the governance of water in the Mekong region requires reconfiguring the 'resources' – policies, norms and material endowments – that enable stakeholders to share power, constituting workable mechanisms of regional governance that includes participatory representation at different scales and a diversity of stakeholders, where nation-state interests recede in favour of regional just water interests.

References

Agarwal, B. (2000) 'Conceptualizing environmental collective action: Why gender matters', *Cambridge Journal of Economics*, vol 24, pp283–310

Ahlers, R. (2005) 'Gender dimensions of neoliberal water policy in Mexico and Bolivia. Empowering or disempowering?', in V. Bennett, S. Davila-Poblete and N. Rico (eds) *Opposing Currents: The Politics of Water and Gender in Latin America*, University of Pittsburgh Press, Pittsburgh, PA, pp53–71

Ahlers, R. and Zwarteveen, M.Z. (2009) 'The water question in feminism: Water control and gender inequities in a neo-liberal era', *Gender, Place and Culture*, vol 16, no 4, pp409–426

Cleaver, F (1999) 'Paradoxes of participation: Questioning participatory approaches to Development', *Journal of International Development*, vol 11, pp597–612

Cleaver, F. and Hamada, K. (2010) '"Good" water governance and gender equity: A troubled relationship', *Gender and Development*, vol 18, no 1, pp27–41

Flyvbjerg, B. (2002) 'Bringing power to planning research: One researcher's praxis story', *Journal of Planning Education and Research*, vol 21, pp353–366

Foran, T., Resurrecion, B., Kansantisukmongkol, C., Wirutskulshai, U., Leeruttanawisut, K. and Lazarus, K. (2010) 'Sustainability assessment of Thailand's electricity planning: Using section 1 of the 2009 Hydropower Sustainability Assessment Protocol', Challenge Program on Water and Food and M-POWER, Vientiane, Laos

Forsyth, T., Walker, A. and Sivaramakrishnan, K. (2008) *Forest Guardians, Forest Destroyers: The Politics of Environmental Knowledge in Northern Thailand*, University of Washington Press, Seattle, WA

Franks, T. and Cleaver, F. (2007) 'Water governance and poverty: A framework for analysis', *Progress in Development Studies*, vol 1, no 4, pp291–306

Fraser, N. (1996) 'Social justice in the age of identity politics: Redistribution, recognition, and participation', Paper for the Tanner Lectures on Human Values, Stanford University, 30 April to 2 May

Fraser, N. (2005) 'Reframing justice in a globalizing world', *New Left Review*, vol 36, pp1–19

Fraser, N. (2008) 'Abnormal justice', *Critical Inquiry*, vol 34, no 3, pp393–422

Granovetter, D. (1992) 'Economic action and social structure: The problem of embeddedness', in Granovetter, M. and Swedburg, R. (eds) *The Sociology of Economic Life*, Westview Press, Oxford

Green, M. (2006) 'Thinking through chronic poverty and destitution: Theorizing social relations and social ordering', Paper presented at a Workshop on Concepts and Methods for Analysing Poverty Dynamics and Chronic Poverty, University of Manchester, UK, 23 to 25 October 2006

Harris, J. (2006) 'Why understanding of social relations matters more for policy on chronic poverty than measurement', Paper presented at a Workshop on Concepts and Methods for Analysing Poverty Dynamics and Chronic Poverty, University of Manchester, UK, 23 to 25 October 2006

Lawrence, S. (2009) 'The Nam Theun 2 controversy and its lessons for Laos', in F. Molle, T. Foran and M. Käkönen (eds) *Contested Waterscapes in the Mekong Region: Hydropower, Livelihoods and Governance*, Earthscan, London

Lebel, L., Garden, P and Imamura, M. (2005) 'Politics of scale, position and place in the governance of water resources in the Mekong region', *Ecology and Society*, vol 10, no 18

Middleton, C., Garcia, J. and Foran, T. (2009) 'Old and new hydropower players in the Mekong Region: Agendas and strategies', on F. Molle, T. Foran and M. Käkönen (eds)

Contested Waterscapes in the Mekong Region: Hydropower, Livelihoods and Governance, Earthscan, London

Ostrom, E. (1990) *Governing the Commons: The Evolution of Institutions for Collective Action*, Cambridge University Press, Cambridge

Roth, D., Zwarteveen, M. and Boelens, R. (2005) *Liquid Relations: Contested Water Rights and Legal Complexity*, Rutgers University Press, New Jersey

Index

access, water resources 10, 245
acid deposition 178
Adaptation Forum 209
adaptive capacity 231
adaptive management, climate change 220
agricultural economy 67
agricultural intensification 206
aluminium melting 174, 175–178
Amu Darya 197
aquaculture
   commercialization 116
   decision-making 131–132
   and fisheries 115, 135–137, 139–140, 151, 157
   natural resources management 133–134
   women 115–116
   work division 121–123
   see also gender relations
aquatic resources 150, 151, 152–155
artisanal villages see craft villages
aset-pasom-pasan (integrated farms) 122, 125, 127, 133
Asian Development Bank 7, 44–45, 209
Asian highlands (Asian water towers) 197–206, 199, 211

Bac Ninh Provincial People's Committee 189
Ban Mae Sa Mai, Thailand 73
Ban Phui Nua, Thailand 75–81, 86
benefits capturing 125–127
biodiversity 198

borders, transcending 228, 248, 250
Brahmaputra 197
Buddhist social norms 128
burdens multiplying, gender relations 127–130
Burma see Myanmar
business-as-usual practices, planning 32

cage culture see aquaculture
Cambodia
   aquaculture 122, 123
   fisheries 120, 123, 124, 126, 128, 129, 132, 134
   gender relations 126, 128, 130, 132
   Greater Mekong Sub-region 209
   Mekong Agreement 208
   Mekong region 2
capture fisheries see fisheries
carbon credits 228
carbon-reduction activities 228
catchphrases see rhetoric
Cham people 130
Champone wetlands, Laos 154
Chao Phraya 2
China
   aquaculture 122–123
   droughts 226
   fisheries 124
   global economic power 209
   Greater Mekong Sub-region 209
   Mekong region 2
   poverty 221
   water resources infrastructure 210
   wetland losses 205

civil society groups 59
clam farming 121
  see also aquaculture
climate
  impacts 217, 219
  risks 222–225
climate change
  adaptive management 220
  effects 200, 202–204
  human development 211
  impacts 223–224
  Mekong region 12, 198, 200, 206–211
  risk governance 230
  risks 217, 219–220, 225–226
  uplands 197–204, 211, 217, 219–220, 225–226
  water availability 226
collaboration 23, 31
collaborative planning 20–22, 26, 30, 31, 32
collective model 174
collective vulnerability 227
commercialization, women 116, 137, 140
common goods, water 189
communicative planning *see* collaborative planning
community-based natural resources management 68
community fisheries 120, 127, 134, 154
compensation, resettlement 53, 55
conflicts *see* resource conflicts; water conflicts
conservation policies 69
coping capacities 233, 235
craft production 168, 169, 171
craft villages 168, 170, 171–172
  see also Man Xa craft village
culture fisheries *see* aquaculture
customary laws 43

dams 24, 151, 155, 210, 221
  see also hydropower; Son La Hydropower Project
deaths, Man Xa 186, 187, 188
debts, household 178
decision-making

access to 8
collaborative planning 30
gender relations 130–133
local people 41, 236
power 6
stakeholders 6, 246
state-centric models 7
villages 78
see also local participation; participation; water governance
Declaration on the Rights of Indigenous Peoples (UNDRIP) 43
development
  and fisheries 150, 160, 161
  and food 150, 151–153, 161
  'good' 161
  national governments 2, 7
  poverty reduction 6, 151, 157
  privileged 245, 246
  and rights 229, 230
  scale 11
  trade-offs 149, 151, 160, 161, 162
  uplands 69, 93, 99, 236
disaster management cycle 236
disaster risk management 237
discrimination 227
diseases 234–235
distributive justice 5, 247
diversity, empowerment 135, 136
division of labour *see* work divisions
doi moi (economic reforms) 39, 167–168, 169, 191
droughts 203, 208, 209, 217, 226, 234

economic growth 210, 245, 246
economic vulnerability 227
elders *see* traditional leaders
electricity 106, 210
Electricity Generating Authority of Thailand (EGAT) 106
El Niño 226
empowerment, women 135–137, 139
energy demand 210
environmental impact assessments 237
environmental law 189
environmentally friendly enterprises 237

environmental narratives 70
environmental protection 42, 168, 170, 190
equity, and social justice 4–5
ethnic communities
    participation 42–44, 57–58
    uplands 208, 227
    Vietnam 40, 42–44, 47, 55, 57–58
    watershed discourse 95–96
    see also Hmong people; Karen people
exclusion, social 229, 246, 251
extreme weather events 201, 203, 205, 226

family units 128, 169
    see also households
famine 158–159
farms
    craft enterprises 170
    integrated 122, 125, 127, 133
    upland 91, 98, 109
    see also aquaculture
FDI (foreign direct investment) enterprises 169
fertilizers 206
financial institutions 7
fish 152–153
fisheries
    and aquaculture 115, 135–137, 139–140, 151, 157
    commercialization 116
    community 120, 127, 134, 154
    crisis narrative 150, 161
    decision-making 132–133
    and development 150, 160, 161
    food source 149, 150, 152–153
    hydropower development 156–158, 161
    livelihoods 153
    natural resources management 134
    poor people 161
    social justice 161, 162
    women 115–116
    work division 119–121
    see also gender relations
fish farms *see* aquaculture

fish ponds 122–123
see also aquaculture
flash floods 226, 232, 233, 234
floods 209, 217, 226, 232, 233–234, 235
food
    development 150, 151–153, 161
    fish 152–153
    forecasts 149
    global crisis 158, 161
    hydropower 150, 151
    importance of 149–150
    livelihoods perspective 160
    rights-based approaches 159
    starvation 158–159
    trade-offs 160, 161
food rights 150, 158
food security 151–152, 153
food sovereignty 159–160
foreign direct investment (FDI) enterprises 169
forests 69
    Khun Kan watershed 104–105
    local people 69, 70, 74, 76, 84, 228
    reforestation 103
    watershed 76–77
    watershed classification 94
    watershed networks 102, 103
    see also Royal Forest Department
frameworks, disputes 31
free, prior and informed consent (FPIC) 43

Ganges 197
gases, metal recycling 178
gender 11
    see also women
gender differentiation 58
gendered space 120–121
gender relations
    benefits capturing 125–127
    decision-making 130–133
    dimensions of 117
    empowerment 135–137, 139
    natural resources management 133–134
    policy implications 138–140

research areas 137–138
work burdens 127–130
work division 119–124
Giay people 55, 57, 58
glaciers 198, 199, 203
global warming *see* climate change
'good development' 161
governments 2, 7, 25–26, 29, 250
Greater Mekong Sub-region (GMS) 209
growth, economic 210, 245, 246

handicraft households 168, 169, 171
hau zos (Hmong leader) 99–100, 108
Haw people 104
headmen, village 101–102
health issues 185–188, 191
hee kho (Karen leader) 100–101, 108
highland Asia (Asian water towers) 197–206, 199, 211
highland Mekong 206, 208
highlands *see* uplands
Himalayas 199, 201
Hmong Environment Networking Group 72–73
Hmong people
  environmental movement 72–74
  fisheries 126
  history 68
  Khun Kan watershed 104
  local meanings 96
  traditional leaders 99–100
  Upper Mae Hae watershed 98
  village headmen 102
  watershed discourse 69–74
  watershed management 75–84, 85, 86
Hoa Binh hydropower project, Vietnam 44
Ho Chi Minh City, Vietnam 171
home, working at 128–129
Hong Nong Sim, Laos 154
host communities 42, 48
households 128
  analysis units 137
  craft production 168, 169, 171
  debts 178
  decision-making 130, 131

livelihood strategies 153, 171, 190
natural resources-dependent 159
see also gender relations
Huai Sai Khao irrigation system 81–83
Huay Pulati stream 103, 104
human rights 219–220, 227, 229
  law 228–229
hunger 150
hydropower 39, 44, 150, 151, 156–158, 161
see also Son La Hydropower Project

ice 198, 199
IMPECT Association 73
inclusion, social 229
indigenous people 43
  see also ethnic communities; local people
individual vulnerability 226
Indus 197, 199
industrial zone management boards 171
industries 170
inequality 190, 231
information, risks 229–230
inland fisheries *see* fisheries
integrated farms 122, 125, 127, 133
integrated water resources management (IWRM) 20–33, 150
international financial institutions 7
international human rights law 229
international water resources governance 19
investments 7
involuntary resettlement *see* resettlement
Irrawaddy 2, 197
irrigation 24, 80–84, 182–183
IWRM (integrated water resources management) 20–33, 150

junk trading 175
justice *see* social justice

Karen people
  environmental movement 70, 72, 74
  irrigation 81, 82, 83, 84
  Khun Kan watershed 104

traditional leaders 100–101, 108
Upper Mae Hae watershed 98
water management 87
watershed classification 96–97
Khong-Chi-Mun irrigation project, Thailand 24
Khon Kaen stakeholder workshop 27–29, 30–31
Khun Kan watershed, Thailand 104–107, 109
Kinh people 43, 47, 130
kronkaan luang (Royal Highland Development Project) 98

labour, division of *see* work division
La Ha people 57, 58
Lancang-Mekong 2
land-use 204–206, 230
language *see* participatory language; rhetoric
La Niña 226
Lao Loum people 126
Laos
  aquatic resources 152–155
  development priorities 151, 157–158
  fisheries 126, 129, 134
  food 151–153
  Greater Mekong Sub-region 209
  health issues 160
  Lao-Thai water transfer 27, 29
  Mekong Agreement 208
  Mekong region 2
  nutrition 157–158
Lao-Thai water transfer 23–30
law
  customary 43
  environmental 189
  human rights 228–229
Law on Environmental Protection (LEP) 170
Law on Water Resources 170
leaders *see* traditional leaders; village headmen
learning, social 224
Lijiang valley, China 124
Lisu people 104

livelihoods
  benefits capturing 125–127
  craft production 168, 172
  fisheries/aquaculture 115, 116, 153
  food 160
  forests 228
  household strategies 7, 93, 153, 171, 190
  Man Xa 178–180
  natural resources 2, 8, 13
  security 6
  work burdens 127–130
  work division 119–124
  see also gender relations; resettlement; *specific pursuits*; work
local adaptive capacity 211
local government 105–107
local participation
  barriers to 40
  communication 52–55
  ethnic minorities 42–44, 57–58
  existing forms 45
  power 6
local people
  aquatic resources 155
  decision-making 41, 236
  diversity 7
  forests 69, 70, 74, 76, 84, 228
  resettlement 40, 41–42
  rights 10
  World Bank 44–45
local politics 91, 92–95

Mae Hae Network Committee (MHNC) 101, 102–104
Mae Sa Nga Watershed Unit (MSN-WU) 102, 103
malaria 234
Man Xa craft village (Vietnam) 174–188
  drinking water 181–182
  health issues 185–188
  history 174–175
  irrigation canal water 182–183
  livelihoods 178–180
  metal recycling 175–178
  public–private interface 188–191

rainwater 173–174, 181–182, 183
sewage ponds 183–184
Van Mon commune 172
wastewater 176, 177–178, 181, 185
water resources degradation 180–181
marginalized people 8
　climate burden 217
　fisheries 150
　participation 57
　risks 219
　uplands 69, 208, 227
　water rights 4, 5
　see also ethnic communities; local
　　people; poor people; upland
　　people; women
Master Plan for the Development of the
　Highland Community, Environment
　and Narcotics Control 96
MDG (Millennium Development Goals)
　150
mechanism, water governance 249–250
Mekong Agreement 208
Mekong Delta 210
Mekong Panel on Climate Change 209
Mekong region
　climate change 198, 200, 207
　climate change implications 201, 206,
　　208–211
　overview 1–2
　uplands 206, 208
Mekong River
　climate change 200
　dams 151, 208
　hydropower 156
　source 197
Mekong River Basin 25, 209
Mekong River Commission (MRC) 208,
　209
Mekong Water Resources Assistance
　Strategy (MWRAS) 24–30
men see gender differentiation; gender
　relations
metal recycling 175–178
MHNC (Mae Hae Network Committee)
　101, 102–104
micro-enterprise production 169

migration, seasonal 235–236
Millennium Development Goals (MDG)
　150
MMSEA (Montane Mainland Southeast
　Asia) 206, 217, 221
mobility 129–130, 135, 136, 230
monsoons 202, 205, 217, 226
Montane Mainland Southeast Asia
　(MMSEA) 206, 217, 221
mountain areas see uplands
MRC (Mekong River Commission) 208,
　209
MSN-WU (Mae Sa Nga Watershed Unit)
　102, 103
MWRAS (Mekong Water Resources
　Assistance Strategy) 24–30
Myanmar 2, 209, 210

Nam Ngum River, Laos 24, 27
Nam Songkhram River, Thailand 131
National Economic and Social
　Development Plans 22–23, 95–96
national food security policies 151–152
National Forest Reserve Act (1964) 94
National Growth and Poverty Eradication
　Strategy (NGPES) 151
national parks 95, 104–105, 106, 107
National Program for Clean Water and
　Environmental Sanitation 170
National Water Vision Statement 22
natural resources management 68,
　133–134, 153
natural resources planning 21
natural water-induced risks 203–204
negotiation spaces 6
neoliberal policies 250
NGPES (National Growth and Poverty
　Eradication Strategy) 151
Ngu Huyen Khue River, Vietnam 174,
　182, 189
nitrogen-based fertilizers 206
Niwat Hongphan 107
novelty, empowerment 135–136
NSEDP (Sixth National Socio Economic
　Development Plan) 151–152
ntoo xeeb trees 73, 74, 77, 78, 79, 80

Nu-Salween 2
nutrition *see* food

organizational cultures 139
outcomes, water governance 250

Pa Mong dam, Thailand 24
participation
    constraints 249
    marginalized people 57
    planning 20
    poor people 6, 251
    term 9, 23
    Vietnam 40–45
    see also collaborative planning; local
        participation
participatory language 20–21, 23, 246
participatory planning *see* collaborative
    planning
Ping River, Thailand 122, 125, 126, 127,
    128–129, 131, 132
place-based inequality 231
planning
    changes in theory 21
    contemporary analysis 19
    participation in 20, 23
    power 21, 30, 32, 251
plantation schemes 228
politics, local 91, 92–95
pollution, water 169, 170, 171, 183, 189
poor people
    causes 160
    diversity 7
    fisheries 161
    food crisis 158
    hydropower development 156
    participation 6, 251
    risks 219
    uplands 208, 221, 228
    vulnerability 219, 230
    see also poverty
poppy crop replacement 67, 68, 75, 76,
    80, 93, 98, 99
population 2, 206, 221, 246
poverty
    causes 221

definitions 150
development rights 230
gender relations 139
Millennium Development Goals 150
vulnerability 231
see also poor people
poverty alleviation
    development 6, 151, 157
    factors associated with 160
    food 150, 160
    indicators 1
    understanding of 248
power
    decision-making 6
    planning 21, 30, 32, 251
precipitation, Asian highlands 201
private sector development 168, 169–
    170, 246
proximity, empowerment 135, 136
public good 190
public–private interface, Man Xa
    188–191

rainwater, Van Mon commune 173–174,
    181–182, 183
recognition, social differences 248
recycling 172, 175–178
redistribution 248
redistributive justice 247
Red River 2
reforestation 103
representation 248
resettlement
    compensation 53, 55
    limitations of 44–45
    local participation 42–44
    local people 40, 41–42
    policy adjustments 47–49
    supervision of 45
    variations 49–51
resource access, multiple scales 84–86
resource conflicts 86, 100, 102, 103, 107,
    108
    see also water conflicts
resource rights 229
resources

competition 11
  societal use of 204–206
  water governance 249, 250
retributive justice 5
RFD *see* Royal Forest Department
rhetoric 20, 26, 32, 246
rice production 152, 154, 198
rights
  and development 229, 230
  food 150, 158, 159
  resource 229
  and risks 220, 230–235
  see also human rights; water rights
rights-based approaches 68, 70–71, 85, 159, 227–230
risk management 235–236
risks
  climate 222–225
  climate change 217, 219–220, 225–226
  definition 219, 222
  information 229–230
  natural water-induced 203–204
  and rights 220, 230–235
risk triangle 222–223
rites-based approaches 68–69, 72–74, 85–86
rivers 2, 197, 199, 203
Royal Forest Department (RFD) 69, 71, 77
  field units 102
  Khun Kan watershed 107
  Royal Projects 99
  watershed classification 94
Royal Highland Development Project (Royal Project; RP) 98, 99
rubber 204, 205

sacred tree practices 73, 74, 77, 78, 79, 80
Saigon River 171
Salween 197, 210
Samik government 29
sanitation 185
Sanyu Consulting 24
Savannakhet province, Laos 154

scale
  development 11
  resource access 84–86
  water governance 250
schistosomiasis 234
sea level rise 198
SEA (Strategic Environmental Assessments) 42, 156
Sen ban ceremony 43
sewage ponds 183–184
Shan people 104
shrimp farming 121–122, 129
  see also aquaculture
Sixth National Socio Economic Development Plan (NSEDP) 151–152
smallholders 159
snow 198, 199
social differences 10, 11, 246, 248
social exclusion 229, 246, 251
social inclusion 229
social justice
  and equity 4–5
  fisheries 161, 162
  framing 247–249
  water governance 13
social learning 224
socially constructed vulnerability 231, 233
social stratification 190
social vulnerability 227
socio-economic inequality 190
Songkhla Lake, Thailand 121
Son La Dam *see* Son La Hydropower Project
Son La Hydropower Project (Vietnam) 46–59
  challenges 46–47
  environmental issues 42
  local participation 51–59, 60
  rationale 46
  resettlement policy 47–49
  resettlement variations 49–51
  surveys 40
space, gendered 120–121
spirit trees 73, 74, 77, 78, 79, 80

stakeholders 6, 20, 27–30, 31, 246
starvation 158–159
state-centric decision-making 7
Strategic Environmental Assessments (SEA) 42, 156
stratification, social 190
sustainable development 20
Syr Darya 197

Tambon Administration Organization (TAO) 105–107
Tam Giang-Cau Hai lagoon, Vietnam 123
Tawee Damrongkiripai 106
Technical Advisory Body for Fisheries Management of the Mekong River Commission (MRC-TAB) 139
temperature, Asian highlands 200
Thac Ba reservoir, Vietnam 44
Thailand 22–33
  agricultural economy 67
  aquaculture 122, 124, 125, 126, 127, 128–129, 131–132
  fisheries 121, 124, 127
  Greater Mekong Sub-region 209
  integrated farms 122, 125, 127, 133
  Khun Kan watershed 104–107, 109
  Lao-Thai water transfer 23–30
  local government 105–107
  Mekong Agreement 208
  Mekong region 2
  national economic and social development plans 22–23
  natural resource management 133
  poverty 221
  Thaksin government 25–26
  upland watersheds contestations 92–95
  watershed discourse 69–74, 95–96
  watershed management 68–69, 75–87
  work burdens 128
  see also Upper Mae Hae Watershed
Thai Leu people 104
Thai people 43, 46, 57, 58
Thaksin government 25–26
Tibetan Plateau 198, 200, 201, 225
Tonle Sap Lake, Cambodia 120, 126, 132, 134, 210

trade-offs
  dam developments 151
  development 149, 151, 160, 161, 162
  environmental protection 168
traditional leaders 98–102, 108
transhumance 235
trees, spirit 73, 74, 77, 78, 79, 80
tuberculosis 230

UNDRIP (Declaration on the Rights of Indigenous Peoples) 43
upland people
  customary laws 43
  downstream dynamics 219, 246
  livelihoods 93
  marginalized 69, 208, 227
  population 221
  poverty 208, 221, 228
  rights-based approaches 227–230
  vulnerability 219
  see also Hmong people; Karen people; local participation; local people
uplands 217–237
  Asia 197, 197–206, 199, 211
  climate change 197–204, 211, 217, 219–220, 225–226
  climate risks 222–225
  development 69, 93, 99, 236
  diversity of 219
  ecological health 211
  farms 91, 98, 109
  land-use change 204–206
  local views of 96–97
  Mekong region 206, 208
  overview 221, 233
  right-based approach 226–230
  risk management 235–236
  risks-rights nexus 230–235
  rivers 197, 199
  value of 236–237
  Vietnam 43
  Water Towers of Asia 197, 199
upland watershed management
  contested watersheds 92–95
  governance 108–110

Khun Kan watershed 104–107
local leaders 98–102
local politics 91
local practices 96
watershed discourse 95–96, 109
watershed terminology 96–97, 109
see also Upper Mae Hae watershed
Upper Mae Hae watershed (Thailand) 97–104
governance 109
local leaders 98–102, 108
Mae Hae Network Committee 101, 102–104

Van Mon commune, Vietnam 172–174
see also Man Xa craft village
Vietnam 39–60
aquaculture 121, 122, 123, 125–126, 129, 130
climate change 198
economic reforms (*doi moi*) 39, 167–168, 169, 191
environmental concerns 167–168
fisheries 123
gender relations 132, 133
Greater Mekong Sub-region 209
Mekong Agreement 208
Mekong region 2
participation 40–45
pollution 170, 183
poverty 221
private sector development 168, 169–170
resettlement 41–42, 43–44
work burdens 128
see also craft villages
Vietnam Rivers Network (VRN) 39
Vietnam Women's Union 139
village headmen 101–102, 108
VRN (Vietnam Rivers Network) 39
vulnerability 225, 226–227
poor people 219, 230, 231
risks-rights nexus 230, 235
socially constructed 231, 233
upland people 219
vulnerability analysis 231

warning systems 236
wastewater
craft villages 170
decentralized management 191
Ho Chi Minh City 171
Man Xa 176, 177–178, 181, 185
water, Asian highlands 198–200
water availability, climate change 226
waterborne diseases 234–235
water conflicts
Huay Pulati stream 103, 104
local leaders 100, 101, 102
Upper Mae Hae watershed 98
watershed networks 102, 103
water governance 2–3
definition 5
improving 249–251
international 19
justice 13
upland 108–110
and water rights 5–8
see also decision-making
water injustice 248
water justice 13
water pollution 169, 170, 171, 183, 189
water resources
access 245
climate change 203
competition 11, 246
degradation 180–181
development 2
development projects 7
land-use change 204–206
unfair allocation 246
see also water use
water resources infrastructure 210, 245
see also dams; hydropower
water resources management
common goods 189, 191
future challenges 209
IWRM 20, 20–22, 22–33, 150
policy framework 170
Thailand 22
trade-offs 149
water rights 3–8, 10
water scarcity 98

watershed management 68–69, 75–87
    see also upland watershed management
watershed networks 101, 102–104
watersheds
    classification 93, 94–97
    definitions 92, 95, 96
    discourse 69–74, 81, 84, 95–96
    forests 76–77
    local terminology 96–97
'Water Towers' of Asia 197, 199
    see also uplands
water use
    Asia 206
    demands 11
    economic growth 210
    future challenge 209
    Man Xa 176
    Upper Mae Hae watershed 98
    see also water resources

weather events, extreme 201, 203, 205, 226
wetland losses 205
wild capture fisheries see fisheries
women
    aquaculture 115–116
    commercialization 116, 137, 140
    empowerment 135–137, 139
    fisheries 115–116
    Man Xa 188
    participation 251
    see also gender; gender differentiation; gender relations
work divisions 119–124
World Bank 7, 44–45

Yangtze 197, 210
Yellow River 197
Yunnan province, China see China